# THE
# HAPPINESS
# HYPOTHESIS
## JONATHAN HAIDT

### Putting Ancient Wisdom
### and Philosophy to the
### Test of Modern Science

arrow books

**29**

Arrow Books
20 Vauxhall Bridge Road
London SW1V 2SA

Arrow Books is part of the Penguin Random House group of companies
whose addresses can be found at global.penguinrandomhouse.com

Penguin
Random House
UK

First published in Great Britain by William Heinemann in 2006
First published in paperback by Arrow Books in 2006

www.penguin.co.uk

A CIP catalogue record for this book is available from the British Library.

ISBN 9780099478898

Typeset by SX Composing DTP, Rayleigh, Essex
**Printed and bound in Great Britain by Clays Ltd, Elcograf S.p.A.**

Penguin Random House is committed to a sustainable future for our business,
our readers and our planet. This book is made from Forest Stewardship
Council® certified paper.

for Jayne

# Contents

# Introduction:
# Too Much Wisdom

WHAT SHOULD I DO, how should I live, and whom should I become? Many of us ask such questions, and, modern life being what it is, we don't have to go far to find answers. Wisdom is now so cheap and abundant that it floods over us from calendar pages, tea bags, bottle caps, and mass e-mail messages forwarded by well-meaning friends. We are in a way like residents of Jorge Luis Borges's *Library of Babel*—an infinite library whose books contain every possible string of letters and, therefore, somewhere an explanation of why the library exists and how to use it. But Borges's librarians suspect that they will never find that book amid the miles of nonsense.

Our prospects are better. Few of our potential sources of wisdom are nonsense, and many are entirely true. Yet, because our library is also effectively infinite—no one person can ever read more than a tiny fraction—we face the paradox of abundance: Quantity undermines the quality of our engagement. With such a vast and wonderful library spread out before us, we often skim books or read just the reviews. We might already have encountered the Greatest Idea, the insight that would have transformed us had we savored it, taken it to heart, and worked it into our lives.

This is a book about ten Great Ideas. Each chapter is an attempt to savor one idea that has been discovered by several of the world's civilizations—to question it in light of what we now know from scientific research, and to extract from it the lessons that still apply to our modern lives.

I am a social psychologist. I do experiments to try to figure out one corner of human social life, and my corner is morality and the moral emotions.
I am also a teacher. I teach a large introductory psychology class at the
University of Virginia in which I try to explain the entire field of psychology
in twenty-four lectures. I have to present a thousand research findings on
everything from the structure of the retina to the workings of love, and
then hope that my students will understand and remember it all. As I
struggled with this challenge in my first year of teaching, I realized that
several ideas kept recurring across lectures, and that often these ideas had
been stated eloquently by past thinkers. To summarize the idea that our
emotions, our reactions to events, and some mental illnesses are caused by
the mental filters through which we look at the world, I could not say it any
more concisely than Shakespeare: "There is nothing either good or bad, but
thinking makes it so."[1] I began to use such quotations to help my students
remember the big ideas in psychology, and I began to wonder just how
many such ideas there were.

To find out, I read dozens of works of ancient wisdom, mostly from the
world's three great zones of classical thought: India (for example, the Upanishads, the Bhagavad Gita, the sayings of the Buddha), China (the Analects
of Confucius, the Tao te Ching, the writings of Meng Tzu and other philosophers), and the cultures of the Mediterranean (the Old and New Testaments, the Greek and Roman philosophers, the Koran). I also read a variety
of other works of philosophy and literature from the last five hundred years.
Every time I found a psychological claim—a statement about human nature
or the workings of the mind or heart—I wrote it down. Whenever I found
an idea expressed in several places and times I considered it a possible
Great Idea. But rather than mechanically listing the top ten all-time most
widespread psychological ideas of humankind, I decided that coherence
was more important than frequency. I wanted to write about a set of ideas
that would fit together, build upon each other, and tell a story about how
human beings can find happiness and meaning in life.

Helping people find happiness and meaning is precisely the goal of the
new field of positive psychology,[2] a field in which I have been active,[3] so
this book is in a way about the origins of positive psychology in ancient wisdom and the applications of positive psychology today. Most of the research

I will cover was done by scientists who would not consider themselves positive psychologists. Nonetheless, I have drawn on ten ancient ideas and a great variety of modern research findings to tell the best story I can about the causes of human flourishing, and the obstacles to well being that we place in our own paths.

The story begins with an account of how the human mind works. Not a full account, of course, just two ancient truths that must be understood before you can take advantage of modern psychology to improve your life. The first truth is the foundational idea of this book: The mind is divided into parts that sometimes conflict. Like a rider on the back of an elephant, the conscious, reasoning part of the mind has only limited control of what the elephant does. Nowadays, we know the causes of these divisions, and a few ways to help the rider and the elephant work better as a team. The second idea is Shakespeare's, about how "thinking makes it so." (Or, as Buddha[4] said, "Our life is the creation of our mind.") But we can improve this ancient idea today by explaining why most people's minds have a bias toward seeing threats and engaging in useless worry. We can also do something to change this bias by using three techniques that increase happiness, one ancient and two very new.

The second step in the story is to give an account of our social lives— again, not a complete account, just two truths, widely known but not sufficiently appreciated. One is the Golden Rule. Reciprocity is the most important tool for getting along with people, and I'll show you how you can use it to solve problems in your own life and avoid being exploited by those who use reciprocity against you. However, reciprocity is more than just a tool. It is also a clue about who we humans are and what we need, a clue that will be important for understanding the end of the larger story. The second truth in this part of the story is that we are all, by nature, hypocrites, and this is why it is so hard for us to follow the Golden Rule faithfully. Recent psychological research has uncovered the mental mechanisms that make us so good at seeing the slightest speck in our neighbor's eye, and so bad at seeing the log in our own. If you know what your mind is up to, and why you so easily see the world through a distorting lens of good and evil, you can take steps to reduce your self-righteousness. You can thereby reduce the frequency of conflicts with others who are equally convinced of their righteousness.

At this point in the story, we'll be ready to ask: Where does happiness come from? There are several different "happiness hypotheses." One is that happiness comes from getting what you want, but we all know (and research confirms) that such happiness is short-lived. A more promising hypothesis is that happiness comes from within and cannot be obtained by making the world conform to your desires. This idea was widespread in the ancient world: Buddha in India and the Stoic philosophers in ancient Greece and Rome all counseled people to break their emotional attachments to people and events, which are always unpredictable and uncontrollable, and to cultivate instead an attitude of acceptance. This ancient idea deserves respect, and it is certainly true that changing your mind is usually a more effective response to frustration than is changing the world. However, I will present evidence that this second version of the happiness hypothesis is wrong. Recent research shows that there are some things worth striving for; there are external conditions of life that can make you lastingly happier. One of these conditions is relatedness—the bonds we form, and need to form, with others. I'll present research showing where love comes from, why passionate love always cools, and what kind of love is "true" love. I'll suggest that the happiness hypothesis offered by Buddha and the Stoics should be amended: Happiness comes from within, and happiness comes from without. We need the guidance of both ancient wisdom and modern science to get the balance right.

The next step in this story about flourishing is to look at the conditions of human growth and development. We've all heard that what doesn't kill us makes us stronger, but that is a dangerous oversimplification. Many of the things that don't kill you can damage you for life. Recent research on "posttraumatic growth" reveals when and why people grow from adversity, and what you can do to prepare yourself for trauma, or to cope with it after the fact. We have also all heard repeated urgings to cultivate virtue in ourselves, because virtue is its own reward, but that, too, is an oversimplification. I'll show how concepts of virtue and morality have changed and narrowed over the centuries, and how ancient ideas about virtue and moral development may hold promise for our own age. I'll also show how positive psychology is beginning to deliver on that promise by offering you a way to "diagnose" and develop your own strengths and virtues.

The conclusion of the story is the question of meaning: Why do some people find meaning, purpose, and fulfillment in life, but others do not? I begin with the culturally widespread idea that there is a vertical, spiritual dimension of human existence. Whether it is called nobility, virtue, or divinity, and whether or not God exists, people simply *do* perceive sacredness, holiness, or some ineffable goodness in others, and in nature. I'll present my own research on the moral emotions of disgust, elevation, and awe to explain how this vertical dimension works, and why the dimension is so important for understanding religious fundamentalism, the political culture war, and the human quest for meaning. I'll also consider what people mean when they ask, "What is the meaning of life?" And I'll give an answer to the question—an answer that draws on ancient ideas about having a purpose but that uses very recent research to go beyond these ancient ideas, or any ideas you are likely to have encountered. In doing so, I'll revise the happiness hypothesis one last time. I could state that final version here in a few words, but I could not explain it in this brief introduction without cheapening it. Words of wisdom, the meaning of life, perhaps even the answer sought by Borges's librarians—all of these may wash over us every day, but they can do little for us unless we savor them, engage with them, question them, improve them, and connect them to our lives. That is my goal in this book.

# 1

# The Divided Self

*For what the flesh desires is opposed to the Spirit, and what
the Spirit desires is opposed to the flesh; for these are opposed
to each other, to prevent you from doing what you want.*

— ST. PAUL, GALATIANS 5:17[1]


*If Passion drives, let Reason hold the Reins.*

— BENJAMIN FRANKLIN[2]


I FIRST RODE A HORSE in 1991, in Great Smoky National Park, North Car-
olina. I'd been on rides as a child where some teenager led the horse by a
short rope, but this was the first time it was just me and a horse, no rope. I
wasn't alone—there were eight other people on eight other horses, and
one of the people was a park ranger—so the ride didn't ask much of me.
There was, however, one difficult moment. We were riding along a path on
a steep hillside, two by two, and my horse was on the outside, walking
about three feet from the edge. Then the path turned sharply to the left,
and my horse was heading straight for the edge. I froze. I knew I had to
steer left, but there was another horse to my left and I didn't want to crash
into it. I might have called out for help, or screamed, "Look out!"; but
some part of me preferred the risk of going over the edge to the certainty
of looking stupid. So I just froze. I did nothing at all during the critical five

seconds in which my horse and the horse to my left calmly turned to the left by themselves.

As my panic subsided, I laughed at my ridiculous fear. The horse knew exactly what she was doing. She'd walked this path a hundred times, and she had no more interest in tumbling to her death than I had. She didn't need me to tell her what to do, and, in fact, the few times I tried to tell her what to do she didn't much seem to care. I had gotten it all so wrong because I had spent the previous ten years driving cars, not horses. Cars go over edges unless you tell them not to.

Human thinking depends on metaphor. We understand new or complex things in relation to things we already know.[3] For example, it's hard to think about life in general, but once you apply the metaphor "life is a journey," the metaphor guides you to some conclusions: You should learn the terrain, pick a direction, find some good traveling companions, and enjoy the trip, because there may be nothing at the end of the road. It's also hard to think about the mind, but once you pick a metaphor it will guide your thinking. Throughout recorded history, people have lived with and tried to control animals, and these animals made their way into ancient metaphors. Buddha, for example, compared the mind to a wild elephant:

> In days gone by this mind of mine used to stray wherever selfish desire or lust or pleasure would lead it. Today this mind does not stray and is under the harmony of control, even as a wild elephant is controlled by the trainer.[4]

Plato used a similar metaphor in which the self (or soul) is a chariot, and the calm, rational part of the mind holds the reins. Plato's charioteer had to control two horses:

> The horse that is on the right, or nobler, side is upright in frame and well jointed, with a high neck and a regal nose; . . . he is a lover of honor with modesty and self-control; companion to true glory, he needs no whip, and is guided by verbal commands alone. The other horse is a crooked great jumble of limbs . . . companion to wild boasts and indecency, he is

shaggy around the ears—deaf as a post—and just barely yields to horse-whip and goad combined.[5]

For Plato, some of the emotions and passions are good (for example, the love of honor), and they help pull the self in the right direction, but others are bad (for example, the appetites and lusts). The goal of Platonic education was to help the charioteer gain perfect control over the two horses. Sigmund Freud offered us a related model 2,300 years later.[6] Freud said that the mind is divided into three parts: the ego (the conscious, rational self); the superego (the conscience, a sometimes too rigid commitment to the rules of society); and the id (the desire for pleasure, lots of it, sooner rather than later). The metaphor I use when I lecture on Freud is to think of the mind as a horse and buggy (a Victorian chariot) in which the driver (the ego) struggles frantically to control a hungry, lustful, and disobedient horse (the id) while the driver's father (the superego) sits in the back seat lecturing the driver on what he is doing wrong. For Freud, the goal of psychoanalysis was to escape this pitiful state by strengthening the ego, thus giving it more control over the id and more independence from the superego.

Freud, Plato, and Buddha all lived in worlds full of domesticated animals. They were familiar with the struggle to assert one's will over a creature much larger than the self. But as the twentieth century wore on, cars replaced horses, and technology gave people ever more control over their physical worlds. When people looked for metaphors, they saw the mind as the driver of a car, or as a program running on a computer. It became possible to forget all about Freud's unconscious, and just study the mechanisms of thinking and decision making. That's what social scientists did in the last third of the century: Social psychologists created "information processing" theories to explain everything from prejudice to friendship. Economists created "rational choice" models to explain why people do what they do. The social sciences were uniting under the idea that people are rational agents who set goals and pursue them intelligently by using the information and resources at their disposal.

But then, why do people keep doing such stupid things? Why do they fail to control themselves and continue to do what they know is not good for them? I, for one, can easily muster the willpower to ignore all the desserts on the menu. But if dessert is placed on the table, I can't resist it. I can resolve to focus on a task and not get up until it is done, yet somehow I find myself walking into the kitchen, or procrastinating in other ways. I can resolve to wake up at 6:00 A.M. to write; yet after I have shut off the alarm, my repeated commands to myself to get out of bed have no effect, and I understand what Plato meant when he described the bad horse as "deaf as a post." But it was during some larger life decisions, about dating, that I really began to grasp the extent of my powerlessness. I would know exactly what I should do, yet, even as I was telling my friends that I would do it, a part of me was dimly aware that I was not going to. Feelings of guilt, lust, or fear were often stronger than reasoning. (On the other hand, I was quite good at lecturing friends in similar situations about what was right for them.) The Roman poet Ovid captured my situation perfectly. In *Metamorphoses*, Medea is torn between her love for Jason and her duty to her father. She laments:

> I am dragged along by a strange new force. Desire and reason are pulling in different directions. I see the right way and approve it, but follow the wrong.[7]

Modern theories about rational choice and information processing don't adequately explain weakness of the will. The older metaphors about controlling animals work beautifully. The image that I came up with for myself, as I marveled at my weakness, was that I was a rider on the back of an elephant. I'm holding the reins in my hands, and by pulling one way or the other I can tell the elephant to turn, to stop, or to go. I can direct things, but only when the elephant doesn't have desires of his own. When the elephant really wants to do something, I'm no match for him.

I have used this metaphor to guide my own thinking for ten years, and when I began to write this book I thought the image of a rider on an elephant would be useful in this first chapter, on the divided self. However, the metaphor has turned out to be useful in every chapter of the book. To understand most important ideas in psychology, you need to understand how the mind is divided into parts that sometimes conflict. We assume that there is one person in each body, but in some ways we are each more like a committee whose members have been thrown together to do a job, but who often find themselves working at cross purposes. Our minds are divided in four ways. The fourth is the most important, for it corresponds most closely to the rider and the elephant; but the first three also contribute to our experiences of temptation, weakness, and internal conflict.

## FIRST DIVISION: MIND VS. BODY

We sometimes say that the body has a mind of its own, but the French philosopher Michel de Montaigne went a step further and suggested that each part of the body has its own emotions and its own agenda. Montaigne was most fascinated by the independence of the penis:

> We are right to note the license and disobedience of this member which thrusts itself forward so inopportunely when we do not want it to, and which so inopportunely lets us down when we most need it. It imperiously contests for authority with our will.[8]

Montaigne also noted the ways in which our facial expressions betray our secret thoughts; our hair stands on end; our hearts race; our tongues fail to speak; and our bowels and anal sphincters undergo "dilations and contractions proper to [themselves], independent of our wishes or even opposed to them." Some of these effects, we now know, are caused by the autonomic nervous system—the network of nerves that controls the organs and glands of our bodies, a network that is completely independent of voluntary or intentional control. But the last item on Montaigne's list—the bowels—reflects the operation of a second brain. Our intestines are lined by a vast network of more than 100 million neurons; these handle all the computations needed to run the chemical refinery that processes and extracts nutrients from food.[9] This gut brain is like a regional administrative center that handles stuff the head brain does not need to bother with. You might expect, then, that this gut brain takes its orders from the head brain and does as it is told. But the gut brain possesses a high degree of auton-omy, and it continues to function well even if the vagus nerve, which con-nects the two brains together, is severed.

The gut brain makes its independence known in many ways: It causes ir-ritable bowel syndrome when it "decides" to flush out the intestines. It trig-gers anxiety in the head brain when it detects infections in the gut, leading you to act in more cautious ways that are appropriate when you are sick.[10] And it reacts in unexpected ways to anything that affects its main neuro-transmitters, such as acetylcholine and serotonin. Hence, many of the ini-tial side effects of Prozac and other selective serotonin reuptake inhibitors involve nausea and changes in bowel function. Trying to improve the work-ings of the head brain can directly interfere with those of the gut brain. The independence of the gut brain, combined with the autonomic nature of changes to the genitals, probably contributed to ancient Indian theories in which the abdomen contains the three lower chakras—energy centers cor-responding to the colon/anus, sexual organs, and gut. The gut chakra is even said to be the source of gut feelings and intuitions, that is, ideas that appear to come from somewhere outside one's own mind. When St. Paul lamented the battle of flesh versus Spirit, he was surely referring to some of the same divisions and frustrations that Montaigne experienced.

## SECOND DIVISION: LEFT VS. RIGHT

A second division was discovered by accident in the 1960s when a surgeon began cutting people's brains in half. The surgeon, Joe Bogen, had a good reason for doing this: He was trying to help people whose lives were de-stroyed by frequent and massive epileptic seizures. The human brain has two separate hemispheres joined by a large bundle of nerves, the corpus callosum. Seizures always begin at one spot in the brain and spread to the surrounding brain tissue. If a seizure crosses over the corpus callosum, it can spread to the entire brain, causing the person to lose consciousness, fall down, and writhe uncontrollably. Just as a military leader might blow up a bridge to prevent an enemy from crossing it, Bogen wanted to sever the corpus callosum to prevent the seizures from spreading.

At first glance this was an insane tactic. The corpus callosum is the largest single bundle of nerves in the entire body, so it must be doing something important. Indeed it is: It allows the two halves of the brain to communicate and coordinate their activity. Yet research on animals found that, within a few weeks of surgery, the animals were pretty much back to normal. So Bogen took a chance with human patients, and it worked. The intensity of the seizures was greatly reduced.

But was there really no loss of ability? To find out, the surgical team brought in a young psychologist, Michael Gazzaniga, whose job was to look for the after-effects of this "split-brain" surgery. Gazzaniga took advantage of the fact that the brain divides its processing of the world into its two hemispheres—left and right. The left hemisphere takes in information from the right half of the world (that is, it receives nerve transmissions from the right arm and leg, the right ear, and the *left* half of each retina, which receives light from the *right* half of the visual field) and sends out commands to move the limbs on the right side of the body. The right hemisphere is in this respect the left's mirror image, taking in information from the left half of the world and controlling movement on the left side of the body. Nobody knows why the signals cross over in this way in all vertebrates; they just do. But in other respects, the two hemispheres are specialized for different tasks. The left hemisphere is specialized for language processing and analytical tasks. In visual tasks, it is better at noticing details. The right hemisphere is better at processing patterns in space, including that all-important pattern, the face. (This is the origin of popular and oversimplified ideas about artists being "right-brained" and scientists being "left-brained").

Gazzaniga used the brain's division of labor to present information to each half of the brain separately. He asked patients to stare at a spot on a screen, and then flashed a word or a picture of an object just to the right of the spot, or just to the left, so quickly that there was not enough time for the patient to move her gaze. If a picture of a hat was flashed just to the right of the spot, the image would register on the left half of each retina (after the image had passed through the cornea and been inverted), which then sent its neural information back to the visual processing areas in the left hemisphere. Gazzaniga would then ask, "What did you see?" Because the left hemisphere has full language capabilities, the patient would quickly and easily say, "A hat." If the image of the hat was flashed to the left of the spot, however, the image was sent back only to the right hemisphere, which does not control speech. When Gazzaniga asked, "What did you see?", the patient, responding from the left hemisphere, said, "Nothing." But when Gazzaniga asked the patient to use her left hand to point to the correct image on a card showing several images, she would point to the hat. Although the right hemisphere had indeed seen the hat, it did not report verbally on what it had seen because it did not have access to the language centers in the left hemisphere. It was as if a separate intelligence was trapped in the right hemisphere, its only output device the left hand.[11]

When Gazzaniga flashed different pictures to the two hemispheres, things grew weirder. On one occasion he flashed a picture of a chicken claw on the right, and a picture of a house and a car covered in snow on the left. The patient was then shown an array of pictures and asked to point to the one that "goes with" what he had seen. The patient's right hand pointed to a picture of a chicken (which went with the chicken claw the left hemisphere had seen), but the left hand pointed to a picture of a shovel (which went with the snow scene presented to the right hemisphere). When the patient was asked to explain his two responses, he did not say, "I have no idea why my left hand is pointing to a shovel; it must be something you showed my right brain." Instead, the left hemisphere instantly made up a plausible story. The patient said, without any hesitation, "Oh, that's easy. The chicken claw goes with the chicken, and you need a shovel to clean out the chicken shed."[12]

This finding, that people will readily fabricate reasons to explain their own behavior, is called "confabulation." Confabulation is so frequent in work with split-brain patients and other people suffering brain damage that Gazzaniga refers to the language centers on the left side of the brain as the interpreter module, whose job is to give a running commentary on whatever the self is doing, even though the interpreter module has no access to the real causes or motives of the self's behavior. For example, if the word "walk" is flashed to the right hemisphere, the patient might stand up and walk away. When asked why he is getting up, he might say, "I'm going to get a Coke." The interpreter module is good at making up explanations, but not at knowing that it has done so.

Science has made even stranger discoveries. In some split-brain patients, or in others who have suffered damage to the corpus callosum, the right hemisphere seems to be actively fighting with the left hemisphere in a condition known as alien hand syndrome. In these cases, one hand, usually the left, acts of its own accord and seems to have its own agenda. The alien hand may pick up a ringing phone, but then refuse to pass the phone to the other hand or bring it up to an ear. The hand rejects choices the person has just made, for example, by putting back on the rack a shirt that the other hand has just picked out. It grabs the wrist of the other hand and tries to stop it from executing the person's conscious plans. Sometimes, the alien hand actually reaches for the person's own neck and tries to strangle him.[13]

These dramatic splits of the mind are caused by rare splits of the brain. Normal people are not split-brained. Yet the split-brain studies were important in psychology because they showed in such an eerie way that the mind is a confederation of modules capable of working independently and even, sometimes, at cross-purposes. Split-brain studies are important for this book because they show in such a dramatic way that one of these modules is good at inventing convincing explanations for your behavior, even when it has no knowledge of the causes of your behavior. Gazzaniga's "interpreter module" is, essentially, the rider. You'll catch the rider confabulating in several later chapters.

## THIRD DIVISION: NEW VS. OLD

If you live in a relatively new suburban house, your home was probably built in less than a year, and its rooms were laid out by an architect who tried to make them fulfill people's needs. The houses on my street, however, were all built around 1900, and since then they have expanded out into their backyards. Porches were extended, then enclosed, then turned into kitchens. Extra bedrooms were built above these extensions, then bathrooms were tacked on to these new rooms. The brain in vertebrates has similarly expanded, but in a forward direction. The brain started off with just three rooms, or clumps of neurons: a hindbrain (connected to the spinal column), a midbrain, and a forebrain (connected to the sensory organs at the front of the animal). Over time, as more complex bodies and behaviors evolved, the brain kept building out the front, away from the spinal column, expanding the forebrain more than any other part. The forebrain of the earliest mammals developed a new outer shell, which included the hypothalamus (specialized to coordinate basic drives and motivations), the hippocampus (specialized for memory), and the amygdala (specialized for emotional learning and responding). These structures are sometimes referred to as the limbic system (from Latin *limbus*, "border" or "margin") because they wrap around the rest of the brain, forming a border.

As mammals grew in size and diversified in behavior (after the dinosaurs became extinct), the remodeling continued. In the more social mammals, particularly among primates, a new layer of neural tissue developed and spread to surround the old limbic system. This neocortex (Latin for "new covering") is the gray matter characteristic of human brains. The front portion of the neocortex is particularly interesting, for parts of it do not appear to be dedicated to specific tasks (such as moving a finger or processing sound). Instead, it is available to make new associations and to engage in thinking, planning, and decision making—mental processes that can free an organism from responding only to an immediate situation.

This growth of the frontal cortex seems like a promising explanation for the divisions we experience in our minds. Perhaps the frontal cortex is the seat of reason: It is Plato's charioteer; it is St. Paul's Spirit. And it has taken over control, though not perfectly, from the more primitive limbic system—Plato's bad horse, St. Paul's flesh. We can call this explanation the Promethean script of human evolution, after the character in Greek mythology who stole fire from the gods and gave it to humans. In this script, our ancestors were mere animals governed by the primitive emotions and drives of the limbic system until they received the divine gift of reason, installed in the newly expanded neocortex.

The Promethean script is pleasing in that it neatly raises us above all other animals, justifying our superiority by our rationality. At the same time, it captures our sense that we are not yet gods—that the fire of rationality is somehow new to us, and we have not yet fully mastered it. The Promethean script also fits well with some important early findings about the roles of the limbic system and the frontal cortex. For example, when some regions of the hypothalamus are stimulated directly with a small electric current, rats, cats, and other mammals can be made gluttonous, ferocious, or hypersexual, suggesting that the limbic system underlies many of our basic animal instincts.[14] Conversely, when people suffer damage to the frontal cortex, they sometimes show an increase in sexual and aggressive behavior because the frontal cortex plays an important role in suppressing or inhibiting behavioral impulses.

There was recently such a case at the University of Virginia's hospital.[15] A schoolteacher in his forties had, fairly suddenly, begun to visit prostitutes, surf child pornography Web sites, and proposition young girls. He was soon arrested and convicted of child molestation. The day before his sentencing, he went to the hospital emergency room because he had a pounding headache and was experiencing a constant urge to rape his landlady. (His wife had thrown him out of the house months earlier.) Even while he was talking to the doctor, he asked passing nurses to sleep with him. A brain scan found that an enormous tumor in his frontal cortex was squeezing everything else, preventing the frontal cortex from doing its job of inhibiting inappropriate behavior and thinking about consequences. (Who in his right mind would put on such a show the day before his sentencing?) When the tumor was removed, the hypersexuality vanished. Moreover, when the tumor grew back the following year, the symptoms returned; and when the tumor was removed again, the symptoms disappeared again.

There is, however, a flaw in the Promethean script: It assumes that reason was installed in the frontal cortex but that emotion stayed behind in the limbic system. In fact, the frontal cortex enabled a great expansion of emotionality in humans. The lower third of the prefrontal cortex is called the orbitofrontal cortex because it is the part of the brain just above the eyes (*orbit* is the Latin term for the eye socket). This region of the cortex has grown especially large in humans and other primates and is one of the most consistently active areas of the brain during emotional reactions.[16] The orbitofrontal cortex plays a central role when you size up the reward and punishment possibilities of a situation; the neurons in this part of the cortex fire wildly when there is an immediate possibility of pleasure or pain, loss or gain.[17] When you feel yourself drawn to a meal, a landscape, or an attractive person, or repelled by a dead animal, a bad song, or a blind date, your orbitofrontal cortex is working hard to give you an emotional feeling of *wanting* to approach or to get away.[18] The orbitofrontal cortex therefore appears to be a better candidate for the id, or for St. Paul's flesh, than for the superego or the Spirit.

The importance of the orbitofrontal cortex for emotion has been further demonstrated by research on brain damage. The neurologist Antonio Damasio has studied people who, because of a stroke, tumor, or blow to the head, have lost various parts of their frontal cortex. In the 1990s, Damasio found that when certain parts of the orbitofrontal cortex are damaged, patients lose most of their emotional lives. They report that when they ought to feel emotion, they feel nothing, and studies of their autonomic reactions (such as those used in lie detector tests) confirm that they lack the normal flashes of bodily reaction that the rest of us experience when observing scenes of horror or beauty. Yet their reasoning and logical abilities are intact. They perform normally on tests of intelligence and knowledge of social rules and moral principles.[19]

So what happens when these people go out into the world? Now that they are free of the distractions of emotion, do they become hyperlogical, able to see through the haze of feelings that blinds the rest of us to the path of perfect rationality? Just the opposite. They find themselves unable to make simple decisions or to set goals, and their lives fall apart. When they look out at the world and think, "What should I do now?" they see dozens of choices but lack immediate internal feelings of like or dislike. They must examine the pros and cons of every choice with their reasoning, but in the absence of feeling they see little reason to pick one or the other. When the rest of us look out at the world, our emotional brains have instantly and automatically appraised the possibilities. One possibility usually jumps out at us as the obvious best one. We need only use reason to weigh the pros and cons when two or three possibilities seem equally good.

Human rationality depends critically on sophisticated emotionality. It is only because our emotional brains works so well that our reasoning can work at all. Plato's image of reason as charioteer controlling the dumb beasts of passion may overstate not only the wisdom but also the power of the charioteer. The metaphor of a rider on an elephant fits Damasio's findings more closely: Reason and emotion must both work together to create intelligent behavior, but emotion (a major part of the elephant) does most of the work. When the neocortex came along, it made the rider possible, but it made the elephant much smarter, too.

## FOURTH DIVISION: CONTROLLED VS. AUTOMATIC

In the 1990s, while I was developing the elephant/rider metaphor for myself, the field of social psychology was coming to a similar view of the mind. After its long infatuation with information processing models and computer metaphors, psychologists began to realize that there are really two processing systems at work in the mind at all times: controlled processes and automatic processes.

Suppose you volunteered to be a subject in the following experiment.[20] First, the experimenter hands you some word problems and tells you to come and get her when you are finished. The word problems are easy: Just unscramble sets of five words and make sentences using four of them. For example, "they her bother see usually" becomes either "they usually see her" or "they usually bother her." A few minutes later, when you have finished the test, you go out to the hallway as instructed. The experimenter is there, but she's engaged in a conversation with someone and isn't making eye contact with you. What do you suppose you'll do? Well, if half the sentences you unscrambled contained words related to rudeness (such as bother, brazen, aggressively), you will probably interrupt the experimenter within a minute or two to say, "Hey, I'm finished. What should I do now?" But if you unscrambled sentences in which the rude words were swapped with words related to politeness ("they her *respect* see usually"), the odds are you'll just sit there meekly and wait until the experimenter acknowledges you—ten minutes from now.

Likewise, exposure to words related to the elderly makes people walk more slowly; words related to professors make people smarter at the game of Trivial Pursuit; and words related to soccer hooligans make people dumber.[21] And these effects don't even depend on your consciously reading the words; the same effects can occur when the words are presented subliminally, that is, flashed on a screen for just a few hundredths of a second, too fast for your conscious mind to register them. But some part of the mind does see the words, and it sets in motion behaviors that psychologists can measure.

According to John Bargh, the pioneer in this research, these experiments show that most mental processes happen automatically, without the need for conscious attention or control. Most automatic processes are completely unconscious, although some of them show a part of themselves to consciousness; for example, we are aware of the "stream of consciousness"[22] that seems to flow on by, following its own rules of association, without any feeling of effort or direction from the self. Bargh contrasts automatic processes with controlled processes, the kind of thinking that takes some effort, that proceeds in steps and that always plays out on the center stage of consciousness. For example, at what time would you need to leave your house to catch a 6:26 flight to London? That's something you have to think about consciously, first choosing a means of transport to the airport and then considering rush-hour traffic, weather, and the strictness of the shoe police at the airport. You can't depart on a hunch. But if you drive to the airport, almost everything you do on the way will be automatic: breathing, blinking, shifting in your seat, daydreaming, keeping enough distance between you and the car in front of you, even scowling and cursing slower drivers.

Controlled processing is limited—we can think consciously about one thing at a time only—but automatic processes run in parallel and can handle many tasks at once. If the mind performs hundreds of operations each second, all but one of them must be handled automatically. So what is the relationship between controlled and automatic processing? Is controlled processing the wise boss, king, or CEO handling the most important questions and setting policy with foresight for the dumber automatic processes to carry out? No, that would bring us right back to the Promethean script and divine reason. To dispel the Promethean script once and for all, it will help to go back in time and look at why we have these two processes, why we have a small rider and a large elephant.

When the first clumps of neurons were forming the first brains more than 600 million years ago, these clumps must have conferred some advantage on the organisms that had them because brains have proliferated ever since. Brains are adaptive because they integrate information from various parts of the animal's body to respond quickly and automatically to threats and opportunities in the environment. By the time we reach 3 million years ago, the Earth was full of animals with extraordinarily sophisticated automatic abilities, among them birds that could navigate by star positions, ants that could cooperate to fight wars and run fungus farms, and several species of hominids that had begun to make tools. Many of these creatures possessed systems of communication, but none of them had developed language.

Controlled processing requires language. You can have bits and pieces of thought through images, but to plan something complex, to weigh the pros and cons of different paths, or to analyze the causes of past successes and failures, you need words. Nobody knows how long ago human beings developed language, but most estimates range from around 2 million years ago, when hominid brains became much bigger, to as recently as 40,000 years ago, the time of cave paintings and other artifacts that reveal unmistakably modern human minds.[23] Whichever end of that range you favor, language, reasoning, and conscious planning arrived in the most recent eye-blink of evolution. They are like new software, Rider version 1.0. The language parts work well, but there are still a lot of bugs in the reasoning and planning programs.[24] Automatic processes, on the other hand, have been through thousands of product cycles and are nearly perfect. This difference in maturity between automatic and controlled processes helps explain why we have inexpensive computers that can solve logic, math, and chess problems better than any human beings can (most of us struggle with these tasks), but none of our robots, no matter how costly, can walk through the woods as well as the average six-year-old child (our perceptual and motor systems are superb).

Evolution never looks ahead. It can't plan the best way to travel from point A to point B. Instead, small changes to existing forms arise (by genetic mutation), and spread within a population to the extent that they help organisms respond more effectively to current conditions. When language evolved, the human brain was not reengineered to hand over the reins of power to the rider (conscious verbal thinking). Things were already working pretty well, and linguistic ability spread to the extent that it helped the elephant do something important in a better way. *The rider evolved to serve to the elephant.* But whatever its origin, once we had it, language was a powerful tool that could be used in new ways, and evolution then selected those individuals who got the best use out of it.

One use of language is that it partially freed humans from "stimulus control." Behaviorists such as B. F. Skinner were able to explain much of the behavior of animals as a set of connections between stimuli and responses. Some of these connections are innate, such as when the sight or smell of an animal's natural food triggers hunger and eating. Other connections are learned, as demonstrated by Ivan Pavlov's dogs, who salivated at the sound of a bell that had earlier announced the arrival of food. The behaviorists saw animals as slaves to their environments and learning histories who blindly respond to the reward properties of whatever they encounter. The behaviorists thought that people were no different from other animals. In this view, St. Paul's lament could be restated as: "My flesh is under stimulus control." It is no accident that we find the carnal pleasures so rewarding. Our brains, like rat brains, are wired so that food and sex give us little bursts of dopamine, the neurotransmitter that is the brain's way of making us enjoy the activities that are good for the survival of our genes.[25] Plato's "bad" horse plays an important role in pulling us toward these things, which helped our ancestors survive and succeed in becoming our ancestors.

But the behaviorists were not exactly right about people. The controlled system allows people to think about long-term goals and thereby escape the tyranny of the here-and-now, the automatic triggering of temptation by the sight of tempting objects. People can imagine alternatives that are not visually present; they can weigh long-term health risks against present pleasures, and they can learn in conversation about which choices will bring success and prestige. Unfortunately, the behaviorists were not entirely wrong about people, either. For although the controlled system does not conform to behaviorist principles, it also has relatively little power to cause behavior. The automatic system was shaped by natural selection to trigger quick and reliable action, and it includes parts of the brain that make us feel pleasure and pain (such as the orbitofrontal cortex) and that trigger survival-related motivations (such as the hypothalamus). The automatic system has its finger on the dopamine release button. The controlled system, in contrast, is better seen as an advisor. It's a rider placed on the elephant's back to help the elephant make better choices. The rider can see farther into the future, and the rider can learn valuable information by talking to other riders or by reading maps, but the rider cannot order the elephant around against its will. I believe the Scottish philosopher David Hume was closer to the truth than was Plato when he said, "Reason is, and ought only to be the slave of the passions, and can never pretend to any other office than to serve and obey them."[26]

In sum, the rider is an advisor or servant; not a king, president, or charioteer with a firm grip on the reins. The rider is Gazzaniga's interpreter module; it is conscious, controlled thought. The elephant, in contrast, is everything else. The elephant includes the gut feelings, visceral reactions, emotions, and intuitions that comprise much of the automatic system. The elephant and the rider each have their own intelligence, and when they work together well they enable the unique brilliance of human beings. But they don't always work together well. Here are three quirks of daily life that illustrate the sometimes complex relationship between the rider and the elephant.

## FAILURES OF SELF CONTROL

Imagine that it is 1970 and you are a four-year-old child in an experiment being conducted by Walter Mischel at Stanford University. You are brought into a room at your preschool where a nice man gives you toys and plays with you for a while. Then the man asks you, first, whether you like marshmallows (you do), and, then, whether you'd rather have this plate here with one marshmallow or that plate there with two marshmallows (that one, of course). Then the man tells you that he has to go out of the room for a little while, and if you can wait until he comes back, you can have the two marshmallows. If you don't want to wait, you can ring this bell here, and he'll come right back and give you the plate with one; but if you do that, you can't have the two. The man leaves. You stare at the marshmallows. You salivate. You want. You fight your wanting. If you are like most four-year-olds, you can hold out for only a few minutes. Then you ring the bell.

Now let's jump ahead to 1985. Mischel has mailed your parents a questionnaire asking them to report on your personality, your ability to delay gratification and deal with frustration, and your performance on your college entrance exams (the Scholastic Aptitude Test). Your parents return the questionnaire. Mischel discovers that the number of seconds you waited to ring the bell in 1970 predicts not only what your parents say about you as a teenager but also the likelihood that you were admitted to a top university. Children who were able to overcome stimulus control and delay gratification for a few extra minutes in 1970 were better able to resist temptation as teenagers, to focus on their studies, and to control themselves when things didn't go the way they wanted.[27]

What was their secret? A large part of it was strategy—the ways that children used their limited mental control to shift attention. In later studies, Mischel discovered that the successful children were those who looked away from the temptation or were able to think about other enjoyable activities.[28] These thinking skills are an aspect of emotional intelligence—an ability to understand and regulate one's own feelings and desires.[29] An emotionally intelligent person has a skilled rider who knows how to distract and coax the elephant without having to engage in a direct contest of wills.

It's hard for the controlled system to beat the automatic system by willpower alone; like a tired muscle,[30] the former soon wears down and caves in, but the latter runs automatically, effortlessly, and endlessly. Once you understand the power of stimulus control, you can use it to your advantage by changing the stimuli in your environment and avoiding undesirable ones; or, if that's not possible, by filling your consciousness with thoughts about their less tempting aspects. Buddhism, for example, in an effort to break people's carnal attachment to their own (and others') flesh, developed methods of meditating on decaying corpses.[31] By choosing to stare at something that revolts the automatic system, the rider can begin to change what the elephant will want in the future.

## MENTAL INTRUSIONS

Edgar Allan Poe understood the divided mind. In *The Imp of the Perverse*, Poe's protagonist carries out the perfect murder, inherits the dead man's estate, and lives for years in healthy enjoyment of his ill-gotten gains. Whenever thoughts of the murder appear on the fringes of his consciousness, he murmurs to himself, "I am safe." All is well until the day he remodels his mantra to "I am safe—yes—if I be not fool enough to make open confession." With that thought, he comes undone. He tries to suppress the thought of confessing, but the harder he tries, the more insistent the thought becomes. He panics, he starts running, people start chasing him, he blacks out, and, when he returns to his senses, he is told that he has made a full confession.

I love this story, for its title above all else. Whenever I am on a cliff, a rooftop, or a high balcony, the imp of the perverse whispers in my ear, "Jump." It's not a command, it's just a word that pops into my consciousness. When I'm at a dinner party sitting next to someone I respect, the imp works hard to suggest the most inappropriate things I could possibly say. Who or what is the imp? Dan Wegner, one of the most perverse and creative social psychologists, has dragged the imp into the lab and made it confess to being an aspect of automatic processing.

In Wegner's studies, participants are asked to try hard *not* to think about something, such as a white bear, or food, or a stereotype. This is hard to do. More important, the moment one stops trying to suppress a thought, the thought comes flooding in and becomes even harder to banish. In other words, Wegner creates minor obsessions in his lab by instructing people not to obsess. Wegner explains this effect as an "ironic process" of mental control.[32] When controlled processing tries to influence thought ("Don't think about a white bear!"), it sets up an explicit goal. And whenever one pursues a goal, a part of the mind automatically monitors progress, so that it can order corrections or know when success has been achieved. When that goal is an action in the world (such as arriving at the airport on time), this feedback system works well. But when the goal is mental, it backfires. Automatic processes continually check: "Am I not thinking about a white bear?" As the act of monitoring for the absence of the thought introduces the thought, the person must try even harder to divert consciousness. Automatic and controlled processes end up working at cross purposes, firing each other up to ever greater exertions. But because controlled processes tire quickly, eventually the inexhaustible automatic processes run unopposed, conjuring up herds of white bears. Thus, the attempt to remove an unpleasant thought can guarantee it a place on your frequent-play list of mental ruminations.

Now, back to me at that dinner party. My simple thought "don't make a fool of yourself" triggers automatic processes looking for signs of foolishness. I know that it would be stupid to comment on that mole on his forehead, or to say "I love you," or to scream obscenities. And up in consciousness, I become aware of three thoughts: comment on the mole, say "I love you," or scream obscenities. These are not commands, just ideas that pop into my head. Freud based much of his theory of psychoanalysis on such mental intrusions and free associations, and he found they often have sexual or aggressive content. But Wegner's research offers a simpler and more innocent explanation: Automatic processes generate thousands of thoughts and images every day, often through random association. The ones that get stuck are the ones that particularly shock us, the ones we try to suppress or deny. The reason we suppress them is not that we know, deep down, that they're true (although some may be), but that they are scary or shameful. Yet once we have tried and failed to suppress them, they can become the sorts of obsessive thoughts that make us believe in Freudian notions of a dark and evil unconscious mind.

## THE DIFFICULTY OF
## WINNING AN ARGUMENT

Consider the following story:

Julie and Mark are sister and brother. They are traveling together in France on summer vacation from college. One night they are staying alone in a cabin near the beach. They decide that it would be interesting and fun if they tried making love. At the very least, it would be a new experience for each of them. Julie is already taking birth control pills, but Mark uses a condom, too, just to be safe. They both enjoy making love, but decide not to do it again. They keep that night as a special secret, which makes them feel even closer to each other.

Do you think it is acceptable for two consenting adults, who happen to be siblings, to make love? If you are like most people in my studies,[33] you immediately answered no. But how would you justify that judgment? People often reach first for the argument that incestuous sex leads to offspring that suffer genetic abnormalities. When I point out that the siblings used two forms of birth control, however, no one says, "Oh, well, in that case it's okay." Instead, people begin searching for other arguments, for example, "It's going to harm their relationship." When I respond that in this case the sex has made the relationship stronger, people just scratch their heads, frown, and say, "I know it's wrong, I'm just having a hard time explaining why."

The point of these studies is that moral judgment is like aesthetic judgment. When you see a painting, you usually know instantly and automatically whether you like it. If someone asks you to explain your judgment, you confabulate. You don't really know why you think something is beautiful, but your interpreter module (the rider) is skilled at making up reasons, as Gazzaniga found in his split-brain studies. You search for a plausible reason for liking the painting, and you latch on to the first reason that makes sense (maybe something vague about color, or light, or the reflection of the painter in the clown's shiny nose). Moral arguments are much the same: Two people feel strongly about an issue, their feelings come first, and their reasons are invented on the fly, to throw at each other. When you refute a person's argument, does she generally change her mind and agree with you? Of course not, because the argument you defeated was not the cause of her position; it was made up after the judgment was already made.

If you listen closely to moral arguments, you can sometimes hear something surprising: that it is really the elephant holding the reins, guiding the rider. It is the elephant who decides what is good or bad, beautiful or ugly. Gut feelings, intuitions, and snap judgments happen constantly and automatically (as Malcolm Gladwell described in *Blink*),[34] but only the rider can string sentences together and create arguments to give to other people. In moral arguments, the rider goes beyond being just an advisor to the elephant; he becomes a lawyer, fighting in the court of public opinion to persuade others of the elephant's point of view.

This, then, is our situation, lamented by St. Paul, Buddha, Ovid, and so many others. Our minds are loose confederations of parts, but we identify with and pay too much attention to one part: conscious verbal thinking. We are like the proverbial drunken man looking for his car keys under the street light. ("Did you drop them here?" asks the cop. "No" says the man, "I dropped them back there in the alley, but the light is better over here.") Because we can see only one little corner of the mind's vast operation, we are surprised when urges, wishes, and temptations emerge, seemingly from nowhere. We make pronouncements, vows, and resolutions, and then are surprised by our own powerlessness to carry them out. We sometimes fall into the view that we are fighting with our unconscious, our id, or our animal self. But really we are the whole thing. We are the rider, and we are the elephant. Both have their strengths and special skills. The rest of this book is about how complex and partly clueless creatures such as ourselves can get along with each other (chapters 3 and 4), find happiness (chapters 5 and 6), grow psychologically and morally (chapters 7 and 8), and find purpose and meaning in our lives (chapters 9 and 10). But first we have to figure out why the elephant is such a pessimist.

# 2

⬥

# Changing Your Mind

*The whole universe is change and life itself is but what you deem it.*

— MARCUS AURELIUS[1]

*What we are today comes from our thoughts of yesterday, and our present thoughts build our life of tomorrow: our life is the creation of our mind.*

— BUDDHA[2]

THE MOST IMPORTANT IDEA in pop psychology is contained in the two quotations above: Events in the world affect us only through our interpretations of them, so if we can control our interpretations, we can control our world. The best-selling self-help advisor of all time, Dale Carnegie, writing in 1944, called the last eight words of the Aurelius quote "eight words that can transform your life."[3] More recently, on television and the Internet, "Dr. Phil" (Phil McGraw) stated as one of his ten "laws of life": "There is no reality, only perception."[4] Self-help books and seminars sometimes seem to consist of little more than lecturing and hectoring people until they understand this idea and its implications for their lives. It can be inspiring to watch: Often a moment comes when a person consumed by years of resentment, pain, and anger realizes that her father (for example) didn't directly hurt her when he abandoned the family; all he did was move out of the house. His action was morally wrong, but the pain came from her reactions to the event, and if she can change those reactions, she can leave behind twenty years of pain and perhaps even get to know her father. The art of pop psychology is to develop a method (beyond lecturing and hectoring) that guides people to that realization.

This art is old. Consider Anicius Boethius, born to one of the most distinguished Roman families in 480 CE, four years after Rome fell to the Goths. Boethius received the best education available in his day and successfully pursued careers in philosophy and public service. He wrote or translated dozens of works on math, science, logic, and theology, at the same time rising to become consul of Rome (the highest elected office) in 510. He was wealthy, he married well, and his sons went on to become consuls themselves. But in 523, at the peak of his power and fortune, Boethius was accused of treason toward the Ostrogoth King Theodoric for remaining loyal to Rome and its Senate. Condemned by the cowardly Senate he had tried to defend, Boethius was stripped of his wealth and honor, thrown into prison on a remote island, and executed in 524.

To take something "philosophically" means to accept a great misfortune without weeping or even suffering. We use this term in part because of the calmness, self-control, and courage that three ancient philosophers—Socrates, Seneca, and Boethius—showed while they awaited their executions. But in *The Consolation of Philosophy*, which Boethius wrote while in prison, he confessed that at first he was anything but philosophical. He wept and wrote poems about weeping. He cursed injustice, and old age, and the Goddess of Fortune, who had blessed him and then abandoned him.

Then one night, while Boethius is wallowing in his wretchedness, the majestic apparition of Lady Philosophy visits him and proceeds to chide him for his unphilosophical behavior. Lady Philosophy then guides Boethius through reinterpretations that foreshadow modern cognitive therapy (described below). She begins by asking Boethius to think about his relationship with the Goddess of Fortune. Philosophy reminds Boethius that Fortune is fickle, coming and going as she pleases. Boethius took Fortune as his mistress, fully aware of her ways, and she stayed with him for a long time. What right has he now to demand that she be chained to his side? Lady Philosophy presents Fortune's defense:

> Why should I alone be deprived of my rights? The heavens are permitted to grant bright days, then blot them out with dark nights; the year may decorate the face of the earth with flowers and fruits, then make it barren again with clouds and frost; the sea is allowed to invite the sailor with fair weather, then terrify him with storms. Shall I, then, permit man's insatiable cupidity to tie me down to a sameness that is alien to my habits?[5]

Lady Philosophy reframes change as normal and as the right of Fortune. ("The whole universe is change," Aurelius had said.) Boethius was fortunate; now he is not. That is no cause for anger. Rather, he should be grateful that he enjoyed Fortune for so long, and he should be calm now that she has left him: "No man can ever be secure until he has been forsaken by Fortune."[6]

Lady Philosophy tries several other reframing tactics. She points out that his wife, sons, and father are each dearer to him than his own life, and all four still live. She helps him see that adverse fortune is more beneficial than good fortune; the latter only makes men greedy for more, but adversity makes them strong. And she draws Boethius's imagination far up into the heavens so that he can look down on the Earth and see it as a tiny speck on which even tinier people play out their comical and ultimately insignificant ambitions. She gets him to admit that riches and fame bring anxiety and avarice, not peace and happiness. After being shown these new perspectives and having his old assumptions challenged, Boethius is finally prepared to absorb the greatest lesson of all, the lesson Buddha and Aurelius had taught centuries earlier: "Nothing is miserable unless you think it so; and on the other hand, nothing brings happiness unless you are content with it."[7] When he takes this lesson to heart, Boethius frees himself from his mental prison. He regains his composure, writes a book that has comforted people for centuries, and faces his death with dignity.

I don't mean to imply that *The Consolation of Philosophy* is just Roman pop psychology, but it does tell a story of freedom through insight that I would like to question. In the previous chapter, I suggested that our divided self is like a rider on the back of an elephant, and I said that we give far too much importance to the rider—conscious thought. Lady Philosophy, like the pop psychology gurus of today, was working with the rider, guiding him to a moment of cognitive insight and reframing. Yet, if you have ever achieved such dramatic insights into your own life and resolved to change your ways or your outlook, you probably found that, three months later, you were right back where you started. Epiphanies can be life-altering,[8] but most fade in days or weeks. The rider can't just decide to change and then order the elephant to go along with the program. Lasting change can come only by retraining the elephant, and that's hard to do. When pop psychology programs are successful in helping people, which they sometimes are, they succeed not because of the initial moment of insight but because they find ways to alter people's behavior over the following months. They keep people involved with the program long enough to retrain the elephant. This chapter is about why the elephant tends toward worry and pessimism in so many people, and about three tools that the rider can use to retrain it.

## THE LIKE-O-METER

The most important words in the elephant's language are "like" and "dislike," or "approach" and "withdraw." Even the simplest animal must make decisions at every moment: Left or right? Go or stop? Eat or don't eat? Animals with brains complex enough to have emotions make these decisions effortlessly and automatically by having what is sometimes called a "like-o-meter" running in their heads at all times. If a monkey tasting a new fruit feels a sweet sensation, its like-o-meter registers "I like it"; the monkey feels pleasure and bites right in. If the taste is bitter, a flash of displeasure discourages further eating. There's no need for a weighing of pros and cons, or for a reasoning system. Just flashes of pleasure and displeasure.

We humans have a like-o-meter too, and it's always running. Its influence is subtle, but careful experiments show that you have a like-dislike reaction to everything you are experiencing, even if you're not aware of the experience. For example, suppose you are a participant in an experiment on what is known as "affective priming." You sit in front of a computer screen and stare at a dot in the center. Every few seconds, a word is flashed over the dot. All you have to do is tap a key with your left hand if the word means something good or likable (such as garden, hope, fun), or tap a key with your right hand if the word means something bad or dislikable (death, tyranny, boredom). It seems easy, but for some reason you find yourself hesitating for a split second on some of the words. Unbeknownst to you, the computer is also flashing up another word, right on the dot, just for a few hundredths of a second before putting up the target word you're rating. Though these words are presented subliminally (below the level of your awareness), your intuitive system is so fast that it reads and reacts to them with a like-o-meter rating. If the subliminal word is *fear,* it would register negative on your like-o-meter, making you feel a tiny flash of displeasure; and then, a split second later, when you see the word *boredom,* you would more quickly say that boredom is bad. Your negative evaluation of boredom has been facilitated, or "primed," by your tiny flash of negativity toward fear. If, however, the word following *fear* is *garden*, you would take longer to say that garden is good, because of the time it takes for your like-o-meter to shift from bad to good.[9]

The discovery of affective priming in the 1980s opened up a world of indirect measurement in psychology. It became possible to bypass the rider and talk directly to the elephant, and what the elephant has to say is sometimes disturbing. For example, what if, instead of flashing subliminal words, we use photographs of black and white faces? Researchers have found that Americans of all ages, classes, and political affiliations react with a flash of negativity to black faces or to other images and words associated with African-American culture.[10] People who report being unprejudiced against blacks show, on average, a slightly smaller automatic prejudice, but apparently the rider and the elephant each have an opinion. (You can test your own elephant at: www.projectimplicit.com.) Even many African Americans show this implicit prejudice, although others show an implicit preference for black faces and names. On balance, African Americans come out with no implicit bias either way.

One of the most bizarre demonstrations of the like-o-meter in action comes from the work of Brett Pelham,[11] who has discovered that one's like-o-meter is triggered by one's own name. Whenever you see or hear a word that resembles your name, a little flash of pleasure biases you toward thinking the thing is good. So when a man named Dennis is considering a career, he ponders the possibilities: "Lawyer, doctor, banker, dentist . . . dentist . . . something about dentist just *feels* right." And, in fact, people named Dennis or Denise are slightly more likely than people with other names to become dentists. Men named Lawrence and women named Laurie are more likely to become lawyers. Louis and Louise are more likely to move to Louisiana or St. Louis, and George and Georgina are more likely to move to Georgia. The own-name preference even shows up in marriage records: People are slightly more likely to marry people whose names sound like their own, even if the similarity is just sharing a first initial. When Pelham presented his findings to my academic department, I was shocked to realize that most of the married people in the room illustrated his claim: Jerry and Judy, Brian and Bethany, and the winners were me, Jon, and my wife, Jayne.

The unsettling implication of Pelham's work is that the three biggest decisions most of us make—what to do with our lives, where to live, and whom to marry—can all be influenced (even if only slightly) by something as trivial as the sound of a name. Life is indeed what we deem it, but the deeming happens quickly and unconsciously. The elephant reacts instinctively and steers the rider toward a new destination.

## NEGATIVITY BIAS

Clinical psychologists sometimes say that two kinds of people seek therapy: those who need tightening, and those who need loosening. But for every patient seeking help in becoming more organized, self-controlled, and responsible about her future, there is a waiting room full of people hoping to loosen up, lighten up, and worry less about the stupid things they said at yesterday's staff meeting or about the rejection they are sure will follow tomorrow's lunch date. For most people, the elephant sees too many things as bad and not enough as good.

It makes sense. If you were designing the mind of a fish, would you have it respond as strongly to opportunities as to threats? No way. The cost of missing a cue that signals food is low; odds are that there are other fish in the sea, and one mistake won't lead to starvation. The cost of missing the sign of a nearby predator, however, can be catastrophic. Game over, end of the line for those genes. Of course, evolution has no designer, but minds created by natural selection end up looking (to us) as though they were designed because they generally produce behavior that is flexibly adaptive in their ecological niches. (See Steven Pinker[12] on how natural selection designs without a designer.) Some commonalities of animal life even create similarities across species that we might call design principles. One such principle is that *bad is stronger than good*. Responses to threats and unpleasantness are faster, stronger, and harder to inhibit than responses to opportunities and pleasures.

This principle, called "negativity bias,"[13] shows up all over psychology. In marital interactions, it takes at least five good or constructive actions to make up for the damage done by one critical or destructive act.[14] In financial transactions and gambles, the pleasure of gaining a certain amount of money is smaller than the pain of losing the same amount.[15] In evaluating a person's character, people estimate that it would take twenty-five acts of life-saving heroism to make up for one act of murder.[16] When preparing a meal, food is easily contaminated (by a single cockroach antenna), but difficult to purify. Over and over again, psychologists find that the human mind reacts to bad things more quickly, strongly, and persistently than to equivalent good things. We can't just will ourselves to see everything as good because our minds are wired to find and react to threats, violations, and setbacks. As Ben Franklin said: "We are not so sensible of the greatest Health as of the least Sickness."[17]

Here's another candidate for a design principle of animal life: Opposing systems push against each other to reach a balance point, but the balance point is adjustable. When you move your arm, one set of muscles extends it and another contracts it. Both are always slightly tensed, ready for action. Your heart rate and breathing are regulated by an autonomic nervous system composed of two subsystems that push your organs in opposite directions: The sympathetic system prepares your body for "fight or flight" and the parasympathetic system calms you down. Both are active all the time, in different ratios. Your behavior is governed by opposing motivational systems: an approach system, which triggers positive emotions and makes you want to move toward certain things; and a withdrawal system, which triggers negative emotions and makes you want to pull back or avoid other things. Both systems are always active, monitoring the environment, and the two systems can produce opposing motives at the same time[18] (as when you feel ambivalence), but their relative balance determines which way you move. (The "like-o-meter" is a metaphor for this balancing process and its subtle moment-by-moment fluctuations.) The balance can shift in an instant: You are drawn by curiosity to an accident scene, but then recoil in horror when you see the blood that you could not have been surprised to see. You want to talk to a stranger, but you find yourself suddenly paralyzed when you approach that person. The withdrawal system can quickly shoot up to full power,[19] overtaking the slower (and generally weaker) approach system.

One reason the withdrawal system is so quick and compelling is that it gets first crack at all incoming information. All neural impulses from the eyes and ears go first to the thalamus, a kind of central switching station in the brain. From the thalamus, neural impulses are sent out to special sensory processing areas in the cortex; and from those areas, information is relayed to the frontal cortex, where it is integrated with other higher mental processes and your ongoing stream of consciousness. If at the end of this process you become aware of a hissing snake in front of you, you could decide to run away and then order your legs to start moving. But because neural impulses move only at about thirty meters per second, this fairly long path, including decision time, could easily take a second or two. It's easy to see why a neural shortcut would be advantageous, and the amygdala is that shortcut. The amygdala, sitting just under the thalamus, dips into the river of unprocessed information flowing through the thalamus, and it responds to patterns that in the past were associated with danger. The amygdala has a direct connection to the part of the brainstem that activates the fight-or-flight response, and if the amygdala finds a pattern that was part of a previous fear episode (such as the sound of a hiss), it orders the body to red alert.[20]

You have felt this happen. If you have ever thought you were alone in a room and then heard a voice behind you, or if you have ever seen a horror movie in which a knife-wielding maniac jumps into the frame without a musical forewarning, you probably flinched, and your heart rate shot up. Your body reacted with fear (via the quick amygdala path) in the first tenth of a second before you could make sense of the event (via the slower cortical path) in the next nine-tenths of a second. Though the amygdala does process some positive information, the brain has no equivalent "green alert" system to notify you instantly of a delicious meal or a likely mate. Such appraisals can take a second or two. Once again, bad is stronger and faster than good. The elephant reacts before the rider even sees the snake on the path. Although you can tell yourself that you are not afraid of snakes, if your elephant fears them and rears up, you'll still be thrown.

One final point about the amygdala: Not only does it reach down to the brainstem to trigger a response to danger but it reaches up to the frontal cortex to change your thinking. It shifts the entire brain over to a withdrawal orientation. There is a two-way street between emotions and conscious thoughts: Thoughts can cause emotions (as when you reflect on a foolish thing you said), but emotions can also cause thoughts, primarily by raising mental filters that bias subsequent information processing. A flash of fear makes you extra vigilant for additional threats; you look at the world through a filter that interprets ambiguous events as possible dangers. A flash of anger toward someone raises a filter through which you see everything the offending person says or does as a further insult or transgression. Feelings of sadness blind you to all pleasures and opportunities. As one famous depressive put it: "How weary, stale, flat, and unprofitable seem to me all the uses of this world!"[21] So when Shakespeare's Hamlet later offers his own paraphrase of Marcus Aurelius—"There is nothing either good or bad but thinking makes it so"[22]—he is right, but he might have added that his negative emotions are making his thinking make everything bad.

## THE CORTICAL LOTTERY

Hamlet was unlucky. His uncle and his mother conspired to murder his father, the king. But his long and deep depressive reaction to this setback suggests that he was unlucky in another way too: He was by nature a pessimist.

When it comes to explaining personality, it's always true that nature and nurture work together. But it's also true that nature plays a bigger role than most people realize. Consider the identical twin sisters Daphne and Barbara. Raised outside London, they both left school at the age of fourteen, went to work in local government, met their future husbands at the age of sixteen at local town hall dances, suffered miscarriages at the same time, and then each gave birth to two boys and a girl. They feared many of the same things (blood and heights) and exhibited unusual habits (each drank her coffee cold; each developed the habit of pushing up her nose with the palm of the hand, a gesture they both called "squidging"). None of this may surprise you until you learn that separate families had adopted Daphne and Barbara as infants; neither even knew of the other's existence until they were reunited at the age of forty. When they finally did meet, they were wearing almost identical clothing.[23]

Such strings of coincidences are common among identical twins who were separated at birth, but they do not happen among fraternal twins who were similarly separated.[24] On just about every trait that has been studied, identical twins (who share all their genes and spend the same nine months in the same womb) are more similar than same-sex fraternal twins (who share only half their genes and spend the same nine months in the same womb). This finding means that genes make at least some contribution to nearly every trait. Whether the trait is intelligence, extroversion, fearfulness, religiosity, political leaning, liking for jazz, or dislike of spicy foods, identical twins are more similar than fraternal twins, and they are usually almost as similar if they were separated at birth.[25] Genes are not blueprints specifying the structure of a person; they are better thought of as *recipes* for producing a person over many years.[26] Because identical twins are created from the same recipe, their brains end up being fairly similar (though not identical), and these similar brains produce many of the same idiosyncratic behaviors. Fraternal twins, on the other hand, are made from two different recipes that happen to share half their instructions. Fraternal twins don't end up being 50 percent similar to each other; they end up with radically different brains, and therefore radically different personalities—almost as different as people from unrelated families.[27]

Daphne and Barbara came to be known as the "giggle twins." Both have sunny personalities and a habit of bursting into laughter in mid-sentence. They won the cortical lottery—their brains were preconfigured to see good in the world. Other pairs of twins, however, were born to look on the dark side. In fact, happiness is one of the most highly heritable aspects of personality. Twin studies generally show that from 50 percent to 80 percent of all the variance among people in their *average* levels of happiness can be explained by differences in their genes rather than in their life experiences.[28] (Particular episodes of joy or depression, however, must usually be understood by looking at how life events interact with a person's emotional predisposition.)

A person's average or typical level of happiness is that person's "affective style." ("Affect" refers to the felt or experienced part of emotion.) Your affective style reflects the everyday balance of power between your approach system and your withdrawal system, and this balance can be read right from your forehead. It has long been known from studies of brainwaves that most people show an asymmetry: more activity either in the right frontal cortex or in the left frontal cortex. In the late 1980s, Richard Davidson at the University of Wisconsin discovered that these asymmetries correlated with a person's general tendencies to experience positive and negative emotions. People showing more of a certain kind of brainwave coming through the left side of the forehead reported feeling more happiness in their daily lives and less fear, anxiety, and shame than people exhibiting higher activity on the right side. Later research showed that these cortical "lefties" are less subject to depression and recover more quickly from negative experiences.[29] The difference between cortical righties and lefties can be seen even in infants: Ten-month-old babies showing more activity on the right side are more likely to cry when separated briefly from their mothers.[30] And this difference in infancy appears to reflect an aspect of personality that is stable, for most people, all the way through adulthood.[31] Babies who show a lot more activity on the right side of the forehead become toddlers who are more anxious about novel situations; as teenagers, they are more likely to be fearful about dating and social activities; and, finally, as adults, they are more likely to need psychotherapy to loosen up. Having lost out in the cortical lottery, they will struggle all their lives to weaken the grip of an overactive withdrawal system. Once when a friend of mine with a negative affective style was bemoaning her life situation, someone suggested that a move to a different city would suit her well. "No," she said, "I can be unhappy anywhere." She might as well have quoted John Milton's paraphrase of Aurelius: "The mind is its own place, and in itself can make a heaven of hell, a hell of heaven."[32]

## SCAN YOUR BRAIN

Which set of statements is more true of you?
Set A:

- I'm always willing to try something new if I think it will be fun.
- If I see a chance to get something I want I move on it right away.
- When good things happen to me, it affects me strongly.
- I often act on the spur of the moment.

Set B:

- I worry about making mistakes.
- Criticism or scolding hurts me quite a bit.
- I feel worried when I think I have done poorly at something important.
- I have many fears compared to my friends.

People who endorse Set A over Set B have a more approach-oriented style and, on average, show greater cortical activity on the left side of the forehead. People who endorse Set B have a more withdrawal-oriented style and, on average, show greater cortical activity on the right side. (Scale adapted from Carver & White, 1994. Copyright © 1994 by the American Psychological Association. Adapted with permission.)

## How to Change Your Mind

If I had an identical twin brother, he would probably dress badly. I have always hated shopping, and I can recognize only six colors by name. Several times I have resolved to improve my style, and have even acceded to women's requests to take me shopping, but it was no use. Each time I quickly returned to my familiar ways, which were stuck in the early 1980s. I couldn't just decide to change, to become something I'm not, by sheer force of will. Instead, I found a more roundabout way to change: I got married. Now I have a closet full of nice clothes, a few pairings that I have memorized as appropriate choices, and a style consultant who recommends variations.

You can change your affective style too—but again, you can't do it by sheer force of will. You have to do something that will change your repertoire of available thoughts. Here are three of the best methods for doing so: meditation, cognitive therapy, and Prozac. All three are effective because they work on the elephant.

### Meditation

Suppose you read about a pill that you could take once a day to reduce anxiety and increase your contentment. Would you take it? Suppose further that the pill has a great variety of side effects, all of them good: increased self-esteem, empathy, and trust; it even improves memory. Suppose, finally, that the pill is all natural and costs nothing. Now would you take it?

The pill exists. It is meditation.[33] It has been discovered by many religious traditions and was in use in India long before Buddha, but Buddhism brought it into mainstream Western culture. There are many kinds of meditation, but they all have in common a conscious attempt to focus attention in a nonanalytical way.[34] It sounds easy: Sit still (in most forms) and focus awareness only on your breathing, or on a word, or on an image, and let no other words, ideas, or images arise in consciousness. Meditation is, however, extraordinarily difficult at first, and confronting your repeated failures in the first weeks teaches the rider lessons in humility and patience. The goal of meditation is to change automatic thought processes, thereby taming the elephant. And the proof of taming is the breaking of attachments.

My dog Andy has two main attachments, through which he interprets everything that happens in my house: eating meat and not being left alone. If my wife and I stand near the front door, he becomes anxious. If we pick up our keys, open the door, and say, "Be a good boy," his tail, head, and somehow even his hips droop pathetically toward the floor. But if we then say, "Andy, come," he's electrified with joy and shoots past us through the doorway. Andy's fear of being left alone gives him many moments of anxiety throughout the day, a few hours of despair (when he is left alone), and a few minutes of joy (each time his solitude is relieved). Andy's pleasures and pains are determined by the choices my wife and I make. If bad is stronger than good, then Andy suffers more from separation than he benefits from reunion.

Most people have many more attachments than Andy; but, according to Buddhism, human psychology is similar to Andy's in many ways. Because Rachel wants to be respected, she lives in constant vigilance for signs of disrespect, and she aches for days after a possible violation. She may enjoy being treated with respect, but disrespect hurts more on average than respect feels good. Charles wants money and lives in a constant state of vigilance for chances to make it: He loses sleep over fines, losses, or transactions that he thinks did not get him the best possible deal. Once again, losses loom larger than gains, so even if Charles grows steadily wealthier, thoughts about money may on average give him more unhappiness than happiness.

For Buddha, attachments are like a game of roulette in which someone else spins the wheel and the game is rigged: The more you play, the more you lose. The only way to win is to step away from the table. And the only way to step away, to make yourself not react to the ups and downs of life, is to meditate and tame the mind. Although you give up the pleasures of winning, you also give up the larger pains of losing.

In chapter 5 I'll question whether this is really a good tradeoff for most people. For now the important point is that Buddha made a psychological discovery that he and his followers embedded in a philosophy and a religion. They have been generous with it, teaching it to people of all faiths and of no faith. The discovery is that meditation tames and calms the elephant. Meditation done every day for several months can help you reduce substantially the frequency of fearful, negative, and grasping thoughts, thereby improving your affective style. As Buddha said: "When a man knows the solitude of silence, and feels the joy of quietness, he is then free from fear and sin."[35]

## Cognitive Therapy

Meditation is a characteristically Eastern solution to the problems of life. Even before Buddha, the Chinese philosopher Lao Tzu had said that the road to wisdom runs through calm inaction, desireless waiting. Western approaches to problems more typically involve pulling out a tool box and trying to fix what's broken. That was Lady Philosophy's approach with her many arguments and reframing techniques. The toolbox was thoroughly modernized in the 1960s by Aaron Beck.

Beck, a psychiatrist at the University of Pennsylvania, had been trained in the Freudian approach in which "the child is father to the man." Whatever ails you is caused by events in your childhood, and the only way to change yourself now is to dig through repressed memories, come up with a diagnosis, and work through your unresolved conflicts. For depressed patients, however, Beck found little evidence in the scientific literature or in his own clinical practice that this approach was working. The more space he gave them to run through their self-critical thoughts and memories of injustice, the worse they felt. But in the late 1960s, when Beck broke with standard practice and, like Lady Philosophy, questioned the legitimacy of his patients' irrational and self-critical thoughts, the patients often seemed to feel better.

Beck took a chance. He mapped out the distorted thought processes characteristic of depressed people and trained his patients to catch and challenge these thoughts. Beck was scorned by his Freudian colleagues, who thought he was treating the symptoms of depression with Band-Aids while letting the disease rage underneath, but his courage and persistence paid off. He created cognitive therapy,[36] one of the most effective treatments available for depression, anxiety, and many other problems.

As I suggested in the last chapter, we often use reasoning not to find the truth but to invent arguments to support our deep and intuitive beliefs (residing in the elephant). Depressed people are convinced in their hearts of three related beliefs, known as Beck's "cognitive triad" of depression. These are: "I'm no good," "My world is bleak," and "My future is hopeless." A depressed person's mind is filled with automatic thoughts supporting these dysfunctional beliefs, particularly when things goes wrong. The thought distortions were so similar across patients that Beck gave them names. Consider the depressed father whose daughter falls down and bangs her head while he is watching her. He instantly flagellates himself with these thoughts: "I'm a terrible father" (this is called "personalization," or seeing the event as a referendum on the self rather than as a minor medical issue); "Why do I always do such terrible things to my children?" ("overgeneralization" combined with dichotomous "always/never" thinking); "Now she's going to have brain damage" ("magnification"); "Everyone will hate me" ("arbitrary inference," or jumping to a conclusion without evidence).

Depressed people are caught in a feedback loop in which distorted thoughts cause negative feelings, which then distort thinking further. Beck's discovery is that you can break the cycle by changing the thoughts. A big part of cognitive therapy is training clients to catch their thoughts, write them down, name the distortions, and then find alternative and more accurate ways of thinking. Over many weeks, the client's thoughts become more realistic, the feedback loop is broken, and the client's anxiety or depression abates. Cognitive therapy works because it teaches the rider how to train the elephant rather than how to defeat it directly in an argument. On the first day of therapy, the rider doesn't realize that the elephant is controlling him, that the elephant's fears are driving his conscious thoughts. Over time, the client learns to use a set of tools; these include challenging automatic thoughts and engaging in simple tasks, such as going out to buy a newspaper rather than staying in bed all day ruminating. These tasks are often assigned as homework, to be done daily. (The elephant learns best from daily practice; a weekly meeting with a therapist is not enough.) With each reframing, and with each simple task accomplished, the client receives a little reward, a little flash of relief or pleasure. And each flash of pleasure is like a peanut given to an elephant as reinforcement for a new behavior. You can't win a tug of war with an angry or fearful elephant, but you can— by gradual shaping of the sort the behaviorists talked about—change your automatic thoughts and, in the process, your affective style. In fact, many therapists combine cognitive therapy with techniques borrowed directly from behaviorism to create what is now called "cognitive behavioral therapy."

Unlike Freud, Beck tested his theories in controlled experiments. People who underwent cognitive therapy for depression got measurably better; they got better faster than people who were put on a waiting list for therapy; and, at least in some studies, they got better faster than those who received other therapies.[37] When cognitive therapy is done very well it is as effective as drugs such as Prozac for the treatment of depression,[38] and its enormous advantage over Prozac is that when cognitive therapy stops, the benefits usually continue because the elephant has been retrained. Prozac, in contrast, works only for as long as you take it.

I don't mean to suggest that cognitive behavioral therapy is the only psychotherapy that works. Most forms of psychotherapy work to some degree, and in some studies they all seem to work equally well.[39] It comes down to a question of fit: Some people respond better to one therapy than another, and some psychological disorders are more effectively treated by one therapy than another. If you have frequent automatic negative thoughts about yourself, your world, or your future, and if these thoughts contribute to chronic feelings of anxiety or despair, then you might find a good fit with cognitive behavioral therapy.[40]

## Prozac

Marcel Proust wrote that "the only true voyage . . . would be not to visit strange lands but to possess other eyes."[41] In the summer of 1996, I tried on a pair of new eyes when I took Paxil, a cousin of Prozac, for eight weeks. For the first few weeks I had only side effects: some nausea, difficulty sleeping through the night, and a variety of physical sensations that I did not know my body could produce, including a feeling I can describe only by saying that my brain felt dry. But then one day in week five, the world changed color. I woke up one morning and no longer felt anxious about the heavy work load and uncertain prospects of an untenured professor. It was like magic. A set of changes I had wanted to make in myself for years— loosening up, lightening up, accepting my mistakes without dwelling on them—happened overnight. However, Paxil had one devastating side effect for me: It made it hard for me to recall facts and names, even those I knew well. I would greet my students and colleagues, reach for a name to put after "Hi," and be left with "Hi . . . there." I decided that as a professor I needed my memory more than I needed peace of mind, so I stopped taking Paxil. Five weeks later, my memory came back, along with my worries. What remained was a firsthand experience of wearing rose-colored glasses, of seeing the world with new eyes.

Prozac was the first member of a class of drugs known as selective serotonin reuptake inhibitors, or SSRIs. In what follows, I use Prozac to stand for the whole group, the psychological effects of which are nearly identical, and which includes Paxil, Zoloft, Celexa, Lexapro, and others. Many things are not known about Prozac and its cousins—above all, how they work. The name of the drug class tells part of the story: Prozac gets into the synapses (the gaps between neurons), but it is *selective* in affecting only synapses that use *serotonin* as their neurotransmitter. Once in the synapses, Prozac *inhibits* the *reuptake* process—the normal process in which a neuron that has just released serotonin into the synapse then sucks it back up into itself, to be released again at the next neural pulse. The net result is that a brain on Prozac has more serotonin in certain synapses, so those neurons fire more often.

So far Prozac sounds like cocaine, heroin, or any other drug that you might have learned is associated with a specific neurotransmitter. But the increase in serotonin happens within a day of taking Prozac, while the benefits don't appear for four to six weeks. Somehow, the neuron on the other side of the synapse is adapting to the new level of serotonin, and it is from that adaptation process that the benefits probably emerge. Or maybe neural adaptation has nothing to do with it. The other leading theory about Prozac is that it raises the level of a neural growth hormone in the hippocampus, a part of the brain crucial for learning and memory. People who have a negative affective style generally have higher levels of stress hormones in their blood; these hormones, in turn, tend to kill off or prune back some critical cells in the hippocampus, whose job, in part, is to shut off the very stress response that is killing them. So people who have a negative affective style may often suffer minor neural damage to the hippocampus, but this can be repaired in four or five weeks after Prozac triggers the release of the neural growth hormone.[42] Although we don't know *how* Prozac works, we do know that it works: It produces benefits above placebo or no-treatment control groups on an astonishing variety of mental maladies, including depression, generalized anxiety disorder, panic attacks, social phobia, premenstrual dysphoric disorder, some eating disorders, and obsessive compulsive disorder.[43]

Prozac is controversial for at least two reasons. First, it is a shortcut. In most studies, Prozac turns out to be just about as effective as cognitive therapy—sometimes a little more, sometimes a little less—but it's so much *easier* than therapy. No daily homework or difficult new skills; no weekly therapy appointment. If you believe in the Protestant work ethic and the maxim "No pain, no gain," then you might be disturbed by Prozac. Second, Prozac does more than just relieve symptoms; it sometimes changes personality. In *Listening to Prozac*,[44] Peter Kramer presents case studies of his patients whose long-standing depression or anxiety was cured by Prozac, and whose personalities then bloomed—greater self-confidence, greater resilience in the face of setbacks, and more joy, all of which sometimes led to big changes in careers and relationships. These cases conform to an idealized medical narrative: person suffers from lifelong disease; medical breakthrough cures disease; person released from shackles, celebrates new freedom; closing shot of person playing joyously with children; fade to black. But Kramer also tells fascinating stories about people who were not ill, who met no diagnostic category for a mental disorder, and who just had the sorts of neuroses and personality quirks that most people have to some degree—fear of criticism, inability to be happy when not in a relationship, tendency to be too critical and overcontrolling of spouse and children. Like all personality traits, these are hard to change, but they are what talk therapy is designed to address. Therapy can't usually change personality, but it can teach you ways of working around your problematic traits. Yet when Kramer prescribed Prozac, the offending traits went away. Lifelong habits, gone overnight (five weeks after starting Prozac), whereas years of psychotherapy often had done nothing. This is why Kramer coined the term "cosmetic psychopharmacology," for Prozac seemed to promise that psychiatrists could shape and perfect minds just as plastic surgeons shape and perfect bodies.

Does that sound like progress, or like Pandora's box? Before you answer that, answer this: Which of these two phrases rings truest to you: "Be all that you can be" or "This above all, to thine own self be true." Our culture endorses both—relentless self-improvement as well as authenticity—but we often escape the contradiction by framing self-improvement as authenticity. Just as gaining an education means struggling for twelve to twenty years to develop one's intellectual potential, character development ought to involve a lifelong struggle to develop one's moral potential. A nine-year-old child does not stay true to herself by keeping the mind and character of a nine-year-old; she works hard to reach her ideal self, pushed and chauffeured by her parents to endless after-school and weekend classes in piano, religion, art, and athletics. As long as change is gradual and a result of the child's hard work, the child is given the moral credit for the change, and that change is in the service of authenticity. But what if there were a pill that enhanced tennis skills? Or a minor surgical technique for implanting piano virtuosity directly and permanently into the brain? Such a separation of self-improvement from authenticity would make many people recoil in horror.

Horror fascinates me, particularly when there is no victim. I study moral reactions to harmless taboo violations such as consensual incest and private flag desecration. These things just *feel* wrong to most people, even when they can't explain why. (I'll explain why in chapter 9.) My research indicates that a small set of innate moral intuitions guide and constrain the world's many moralities, and one of these intuitions is that the body is a temple housing a soul within.[45] Even people who do not consciously believe in God or the soul are offended by or feel uncomfortable about someone who treats her body like a playground, its sole purpose to provide pleasure. A shy woman who gets a nose job, breast augmentation, twelve body piercings, and a prescription for elective Prozac would be as shocking to many people as a minister who remodels his church to look like an Ottoman harem.

The transformation of the church might hurt others by causing several parishioners to die from apoplexy. It is hard, however, to find harm in the self-transformer beyond some vague notion that she is "not being true to herself." But if this woman had previously been unhappy with her hypersensitive and overly inhibited personality, and if she had made little progress with psychotherapy, why exactly should she be true to a self she doesn't want? Why not change herself for the better? When I took Paxil, it changed my affective style for the better. It made me into something I was not, but had long wanted to be: a person who worries less, and who sees the world as being full of possibilities, not threats. Paxil improved the balance between my approach and withdrawal systems, and had there been no side effects, I would still be taking it today.

I therefore question the widespread view that Prozac and other drugs in its class are overprescribed. It's easy for those who did well in the cortical lottery to preach about the importance of hard work and the unnaturalness of chemical shortcuts. But for those who, through no fault of their own, ended up on the negative half of the affective style spectrum, Prozac is a way to compensate for the unfairness of the cortical lottery. Furthermore, it's easy for those who believe that the body is a temple to say that cosmetic psychopharmacology is a kind of sacrilege. Something is indeed lost when psychiatrists no longer listen to their patients as people, but rather as a car mechanic would listen to an engine, looking only for clues about which knob to adjust next. But if the hippocampal theory of Prozac is correct, many people really do need a mechanical adjustment. It's as though they had been driving for years with the emergency break halfway engaged, and it might be worth a five-week experiment to see what happens to their lives when the brake is released. Framed in this way, Prozac for the "worried well" is no longer just cosmetic. It is more like giving contact lenses to a person with poor but functional eyesight who has learned ways of coping with her limitations. Far from being a betrayal of that person's "true self," contact lenses can be a reasonable shortcut to proper functioning.

The epigraphs that opened this chapter are true. Life is what we deem it, and our lives are the creations of our minds. But these claims are not helpful until augmented by a theory of the divided self (such as the rider and the elephant) and an understanding of negativity bias and affective style. Once you know why change is so hard, you can drop the brute force method and take a more psychologically sophisticated approach to self-improvement. Buddha got it exactly right: You need a method for taming the elephant, for changing your mind gradually. Meditation, cognitive ther-apy, and Prozac are three effective means of doing so. Because each will be effective for some people and not for others, I believe that all three should be readily available and widely publicized. Life itself is but what you deem it, and you can—through meditation, cognitive therapy, and Prozac—redeem yourself.

# 3

## Reciprocity with a Vengeance

> Zigong asked: "Is there any single word that could guide one's
> entire life?" The master said: "Should it not be reciprocity?
> What you do not wish for yourself, do not do to others."
>
> —ANALECTS OF CONFUCIUS[1]

> That which is hateful to you, do not do to your fellow; this, in
> a few words, is the entire Torah; all the rest is but an elabora-
> tion of this one, central point.
>
> —RABBI HILLEL, 1ST CENT. BCE[2]

WHEN THE SAGES PICK a single word or principle to elevate above all others, the winner is almost always either "love" or "reciprocity." Chapter 6
will cover love; this chapter is about reciprocity. Both are, ultimately, about
the same thing: the bonds that tie us to one another.

The opening scene of the movie *The Godfather* is an exquisite portrayal
of reciprocity in action. It is the wedding day of the daughter of the Godfather, Don Corleone. The Italian immigrant Bonasera, an undertaker, has
come to ask for a favor: He wants to avenge an assault upon the honor and
body of his own daughter, who was beaten by her boyfriend and another
young man. Bonasera describes the assault, the arrest, and the trial of the
two boys. The judge gave them a suspended sentence and let them go free

that very day. Bonasera is furious and feels humiliated; he has come to Don Corleone to ask that justice be done. Corleone asks what exactly he wants. Bonasera whispers something into his ear, which we can safely assume is "Kill them." Corleone refuses, and points out that Bonasera has not been much of a friend until now. Bonasera admits he was afraid of getting into "trouble." The dialogue continues:[3]

CORLEONE: I understand. You found paradise in America, you had a good trade, made a good living. The police protected you and there were courts of law. And you didn't need a friend like me. But now you come to me and you say, "Don Corleone give me justice." But you don't ask with respect. You don't offer friendship. You don't even think to call me "Godfather." Instead, you come into my house on the day my daughter is to be married, and you ask me to do murder, for money.

BONASERA: I ask you for justice.

CORLEONE: That is not justice; your daughter is still alive.

BONASERA: Let them suffer then, as she suffers. [Pause]. How much shall I pay you?

CORLEONE: Bonasera . . . Bonasera . . . What have I ever done to make you treat me so disrespectfully? If you'd come to me in friendship, then this scum that ruined your daughter would be suffering this very day. And if by chance an honest man like yourself should make enemies, then they would become *my* enemies. And then they would fear you.

BONASERA: Be my friend—[He bows to Corleone]—Godfather? [He kisses Corleone's hand].

CORLEONE: Good. [Pause.] Some day, and that day may never come, I'll call upon you to do a service for me. But until that day—accept this justice as a gift on my daughter's wedding day.

The scene is extraordinary, a kind of overture that introduces the themes of violence, kinship, and morality that drive the rest of the movie. But just as extraordinary to me is how easy it is for us to understand this complex interaction in an alien subculture. We intuitively understand why Bonasera wants the boys killed, and why Corleone refuses to do it. We wince at

Bonasera's clumsy attempt to offer money when what is lacking is the right relationship, and we understand why Bonasera had been wary, before, of cultivating the right relationship. We understand that in accepting a "gift" from a mafia don, a chain, not just a string, is attached. We understand all of this effortlessly because we see the world through the lens of reciprocity. Reciprocity is a deep instinct; it is the basic currency of social life. Bonasera uses it to buy revenge, which is itself a form of reciprocity. Corleone uses it to manipulate Bonasera into joining Corleone's extended family. In the rest of this chapter I'll explain how we came to adopt reciprocity as our social currency, and how you can spend it wisely.

## ULTRASOCIALITY

Animals that fly seem to violate the laws of physics, but only until you learn a bit more about physics. Flight evolved independently at least three times in the animal kingdom: in insects, dinosaurs (including modern birds), and mammals (bats). In each case, a physical feature that had potentially aerodynamic properties was already present (for example, scales that lengthened into feathers, which later made gliding possible).

Animals that live in large peaceful societies seem to violate the laws of evolution (such as competition and survival of the fittest), but only until you learn a bit more about evolution. Ultrasociality[4]—living in large cooperative societies in which hundreds or thousands of individuals reap the benefits of an extensive division of labor—evolved independently at least four times in the animal kingdom: among hymenoptera (ants, bees, and wasps); termites; naked mole rats; and humans. In each case, a feature possessing potentially cooperation-enhancing properties already existed. For all the nonhuman ultrasocial species, that feature was the genetics of kin altruism. It's obvious that animals will risk their lives for the safety of their own children: The only way to "win" at the game of evolution is to leave surviving copies of your genes. Yet not just your children carry copies of your genes. Your siblings are just as closely related to you (50 percent shared genes) as your children; your nephews and nieces share a quarter of your genes, and your cousins one eighth. In a strictly Darwinian calculation, whatever cost you would bear to save one of your children you should be willing to pay to save two nieces or four cousins.[5]

Because nearly all animals that live in cooperative groups live in groups of close relatives, most altruism in the animal kingdom reflects the simple axiom that shared genes equals shared interests. But because the sharing drops off so quickly with each fork in the family tree (second cousins share only one thirty-second of their genes), kin altruism explains only how groups of a few dozen, or perhaps a hundred, animals can work together. Out of a flock of thousands, only a small percentage would be close enough to be worth taking risks for. The rest would be competitors, in the Darwinian sense. Here's where the ancestors of bees, termites, and mole rats took the common mechanism of kin altruism, which makes many species sociable, and parlayed it[6] into the foundation of their uncommon ultrasociality: They are all siblings. Those species each evolved a reproduction system in which a single queen produces all the children, and nearly all the children are either sterile (ants) or else their reproductive abilities are suppressed (bees, mole rats); therefore, a hive, nest, or colony of these animals is one big family. If everyone around you is your sibling, and if the survival of your genes depends on the survival of your queen, selfishness becomes genetic suicide. These ultrasocial species display levels of cooperation and self-sacrifice that still astonish and inspire those who study them. Some ants, for example, spend their lives hanging from the top of a tunnel, offering their abdomens for use as food storage bags by the rest of the nest.[7]

The ultrasocial animals evolved into a state of ultrakinship, which led automatically to ultracooperation (as in building and defending a large nest or hive), which allowed the massive division of labor (ants have castes such as soldier, forager, nursery worker, and food storage bag), which created hives overflowing with milk and honey, or whatever other substance they use to store their surplus food. We humans also try to extend the reach of kin altruism by using fictitious kinship names for nonrelatives, as when children are encouraged to call their parents' friends Uncle Bob and Aunt Sarah. Indeed, the mafia is known as "the family," and the very idea of a godfather is an attempt to forge a kin-like link with a man who is not true kin. The human mind finds kinship deeply appealing, and kin altruism surely underlies the cultural ubiquity of nepotism. But even in the mafia, kin altruism can take you only so far. At some point you have to work with people who are at best distant relations, and to do so you'd better have another trick up your sleeve.

## You Scratch My Back, I'll Scratch Yours

What would you do if you received a Christmas card from a complete stranger? This actually happened in a study in which a psychologist sent Christmas cards to people at random. The great majority sent him a card in return.[8] In his insightful book *Influence*,[9] Robert Cialdini of Arizona State University cites this and other studies as evidence that people have a mindless, automatic reciprocity reflex. Like other animals, we will perform certain behaviors when the world presents us with certain patterns of input. A baby herring gull, seeing a red spot on its mother's beak, pecks at it automatically, and out comes regurgitated food. The baby gull will peck just as vigorously at a red spot painted on the end of a pencil. A cat stalks a mouse using the same low-down, wiggle-close-then-pounce technique used by cats around the world. The cat uses the same technique to attack a string trailing a ball of yarn because the string accidentally activates the cat's mouse-tail-detector module. Cialdini sees human reciprocity as a similar ethological reflex: a person receives a favor from an acquaintance and wants to repay the favor. The person will even repay an empty favor from a stranger, such as the receipt of a worthless Christmas card.

The animal and human examples are not exactly parallel, however. The gulls and cats are responding to visual stimuli with specific bodily movements, executed immediately. The person is responding to the *meaning* of a situation with a motivation that can be satisfied by a variety of bodily movements executed days later. So what is really built into the person is a *strategy*: Play tit for tat. Do to others what they do unto you. Specifically, the tit-for-tat strategy is to be nice on the first round of interaction; but after that, do to your partner whatever your partner did to you on the previous round.[10] Tit for tat takes us way beyond kin altruism. It opens the possibility of forming cooperative relationships with strangers.

Most interactions among animals (other than close kin) are zero-sum games: One animal's gain is the other's loss. But life is full of situations in which cooperation would expand the pie to be shared if only a way could be found to cooperate without being exploited. Animals that hunt are particularly vulnerable to the variability of success: They may find far more food than they can eat in one day, and then find no food at all for three weeks. Animals that can trade their surplus on a day of plenty for a loan on a day of need are much more likely to survive the vagaries of chance. Vampire bats, for example, will regurgitate blood from a successful night of bloodsucking into the mouth of an unsuccessful and genetically unrelated peer. Such behavior seems to violate the spirit of Darwinian competition, except that the bats keep track of who has helped them in the past, and in return they share primarily with those bats.[11] Like the Godfather, bats play tit for tat, and so do other social animals, particularly those that live in relatively small, stable groups where individuals can recognize each other as individuals.[12]

But if the response to noncooperation is just noncooperation on the next round, then tit for tat can unite groups of only a few hundred. In a large enough group, a cheating vampire bat can beg a meal from a different successful bat each night and, when they come to him pleading for a return favor, just wrap his wings around his head and pretend to be asleep. What are they going to do to him? Well, if these were people rather than bats, we know what they'd do: They'd beat the hell out of him. Vengeance and gratitude are moral sentiments that amplify and enforce tit for tat. Vengeful and grateful feelings appear to have evolved precisely because they are such useful tools for helping individuals create cooperative relationships, thereby reaping the gains from non-zero-sum games.[13] A species equipped with vengeance and gratitude responses can support larger and more cooperative social groups because the payoff to cheaters is reduced by the costs they bear in making enemies.[14] Conversely, the benefits of generosity are increased because one gains friends.

Tit for tat appears to be built into human nature as a set of moral emotions that make us *want* to return favor for favor, insult for insult, tooth for tooth, and eye for eye. Several recent theorists[15] even talk about an "exchange organ" in the human brain, as though a part of the brain were devoted to keeping track of fairness, debts owed, and social accounts-receivable. The "organ" is a metaphor—nobody expects to find an isolated blob of brain tissue the only function of which is to enforce reciprocity. However, recent evidence suggests that there really could be an exchange organ in the brain if we loosen the meaning of "organ" and allow that functional systems in the brain are often composed of widely separated bits of neural tissue that work together to do a specific job.

Suppose you were invited to play the "ultimatum" game, which economists invented[16] to study the tension between fairness and greed. It goes like this: Two people come to the lab but never meet. The experimenter gives one of them—let's suppose it's not you—twenty one-dollar bills and asks her to divide them between the two of you in any way she likes. She then gives you an ultimatum: Take it or leave it. The catch is that if you leave it, if you say no, you both get nothing. If you were both perfectly rational, as most economists would predict, your partner would offer you one dollar, knowing that you'd prefer one dollar to no dollars, and you'd accept her offer, because she was right about you. But the economists were wrong about you both. In real life, nobody offers one dollar, and around half of all people offer ten dollars. But what would you do if your partner offered you seven dollars? Or five? Or three? Most people would accept the seven dollars, but not the three. Most people are willing to pay a few dollars, but not seven, to punish the selfish partner.

Now suppose you played this game while inside an fMRI scanner. Alan Sanfey[17] and his colleagues at Princeton had people do just that; the researchers then looked at what parts of the brain were more active when people were given unfair offers. One of the three areas that differed most (when comparing responses to unfair vs. fair offers) was the frontal insula, an area of the cortex on the frontal underside of the brain. The frontal insula is known to be active during most negative or unpleasant emotional states, particularly anger and disgust. Another area was the dorsolateral prefrontal cortex, just behind the sides of the forehead, known to be active during reasoning and calculation. Perhaps the most impressive finding from Sanfey's study is that people's ultimate response—accept or reject—could be predicted by looking at the state of their brains moments before they pressed a button to make a choice. Those subjects who showed more activation in the insula than in the dorsolateral prefrontal cortex generally went on to reject the unfair offer; those with the reverse pattern generally accepted it. (It's no wonder that marketers, political consultants, and the CIA are so interested in neural imaging and "neuromarketing.")

Gratitude and vengefulness are big steps on the road that led to human ultrasociality, and it's important to realize that they are two sides of one coin. It would be hard to evolve one without the other. An individual who had gratitude without vengefulness would be an easy mark for exploitation, and a vengeful and ungrateful individual would quickly alienate all potential cooperative partners. Gratitude and revenge are also, not coincidentally, major forces holding together the mafia. The Godfather sits at the center of a vast web of reciprocal obligations and favors. He accumulates power with each favor he does, secure in the knowledge that nobody who values his own life will fail to repay at a time of the Godfather's choosing. Revenge for most of us is much less drastic, but if you have worked long enough in an office, restaurant, or store, you know there are many subtle ways to retaliate against those who have crossed you, and many ways to help those who have helped you.

## YOU STAB HIS BACK, I'LL STAB YOURS

When I said that people would beat the hell out of an ingrate who failed to repay an important favor, I left out a qualification. For a first offense, they'd probably just gossip. They'd ruin his reputation. Gossip is another key piece in the puzzle of how humans became ultrasocial. It might also be the reason we have such large heads.

Woody Allen once described his brain as his "second favorite organ," but for all of us it's by far the most expensive one to run. It accounts for 2 percent of our body weight but consumes 20 percent of our energy. Human brains grow so large that human beings must be born prematurely[18] (at least, compared to other mammals, who are born when their brains are more or less ready to control their bodies), and even then they can barely make it through the birth canal. Once out of the womb, these giant brains attached to helpless baby bodies require somebody to carry them around for a year or two. The tripling of human brain size from the time of our last common ancestor with chimpanzees to today imposed tremendous costs on parents, so there must have been a very good reason to do it. Some have argued that the reason was hunting and tool making, others suggest that the extra gray matter helped our ancestors locate fruit. But the only theory that explains why animals in general have particular brain sizes is the one that maps brain size onto social group size. Robin Dunbar[19] has demonstrated that *within* a given group of vertebrate species—primates, carnivores, ungulates, birds, reptiles, or fish—the logarithm of the brain size is almost perfectly proportional to the logarithm of the social group size. In other words, all over the animal kingdom, brains grow to manage larger and larger groups. Social animals are smart animals.

Dunbar points out that chimpanzees live in groups of around thirty, and like all social primates, they spend enormous amounts of time grooming each other. Human beings ought to live in groups of around 150 people, judging from the logarithm of our brain size; and sure enough, studies of hunter-gatherer groups, military units, and city dwellers' address books suggest that 100 to 150 is the "natural" group size within which people can know just about everyone directly, by name and face, and know how each person is related to everybody else. But if grooming is so central to primate sociality, and if our ancestors began living in larger and larger groups (for some other reason, such as to take advantage of a new ecological niche with high predation risks), at some point grooming became an inadequate means of keeping up one's relationships.

Dunbar suggests that language evolved as a replacement for physical grooming.[20] Language allows small groups of people to bond quickly and to learn from each other about the bonds of others. Dunbar notes that people do in fact use language primarily to talk about other people—to find out who is doing what to whom, who is coupling with whom, who is fighting with whom. And Dunbar points out that in our ultrasocial species, success is largely a matter of playing the social game well. It's not what you know, it's who you know. In short, Dunbar proposes that language evolved because it enabled gossip. Individuals who could share social information, using any primitive means of communication, had an advantage over those who could not. And once people began gossiping, there was a runaway competition to master the arts of social manipulation, relationship aggression, and reputation management, all of which require yet more brain power.

Nobody knows how language evolved, but I find Dunbar's speculation so fascinating that I love to tell people about it. It's not good gossip—after all, you don't know Dunbar—but if you are like me you have an urge to tell your friends about anything you learn that amazes or fascinates you, and this urge itself illustrates Dunbar's point: We are *motivated* to pass on information to our friends; we even sometimes say, "I can't keep it in, I have to tell somebody." And when you do pass on a piece of juicy gossip, what happens? Your friend's reciprocity reflex kicks in and she feels a slight pressure to return the favor. If she knows something about the person or event in question, she is likely to speak up: "Oh really? Well, I heard that he . . ." Gossip elicits gossip, and it enables us to keep track of everyone's reputation without having to witness their good and bad deeds personally. Gossip creates a non-zero-sum game because it costs us nothing to give each other information, yet we both benefit by receiving information.

Because I'm particularly interested in the role of gossip in our moral lives, I was pleased when a graduate student in my department, Holly Hom, told me that she wanted to study gossip. In one of Holly's studies,[21] we asked fifty-one people to fill out a short questionnaire each time over the course of a week that they took part in a conversation that went on for at least ten minutes. We then took only the records in which the topic of conversation was another person, which gave us about one episode of potential gossip per day per person. Among our main findings: Gossip is overwhelmingly critical, and it is primarily about the moral and social violations of others. (For college students, this meant a lot of talk about the sexuality, cleanliness, and drinking habits of their friends and roommates.) People do occasionally tell stories about the good deeds of others, but such stories are only one tenth as common as stories about transgressions. When people pass along high-quality ("juicy") gossip, they feel more powerful, they have a better shared sense of what is right and what's wrong, and they feel more closely connected to their gossip partners.

A second study revealed that most people hold negative views of gossip and gossipers, even though almost everyone gossips. When we compared people's attitudes about gossip to the social functions that gossip serves, Holly and I came to believe that gossip is underappreciated. In a world with no gossip, people would not get away with murder but they would get away with a trail of rude, selfish, and antisocial acts, often oblivious to their own violations. Gossip extends our moral–emotional toolkit. In a gossipy world, we don't just feel vengeance and gratitude toward those who hurt or help us; we feel pale but still instructive flashes of contempt and anger toward people whom we might not even know. We feel vicarious shame and embarrassment when we hear about people whose schemes, lusts, and private failings are exposed. Gossip is a policeman and a teacher. Without it, there would be chaos and ignorance.[22]

Many species reciprocate, but only humans gossip, and much of what we gossip about is the value of other people as partners for reciprocal relationships. Using these tools, we create an ultrasocial world, a world in which we refrain from nearly all the ways we could take advantage of those weaker than us, a world in which we often help those who are unlikely ever to be able to return the favor. We *want* to play tit for tat, which means starting out nice without being a pushover, and we *want* to cultivate a reputation for being a good player. Gossip and reputation make sure that what goes around comes around—a person who is cruel will find that others are cruel back to him, and a person who is kind will find that other others are kind in return. Gossip paired with reciprocity allow karma to work here on earth, not in the next life. As long as everyone plays tit-for-tat augmented by gratitude, vengeance, and gossip, the whole system should work beautifully. (It rarely does, however, because of our self-serving biases and massive hypocrisy. See chapter 4.)

## USE THE FORCE, LUKE

In offering reciprocity as the best word to guide one's life, Confucius was wise. Reciprocity is like a magic wand that can clear your way through the jungle of social life. But as anyone who has read a Harry Potter book knows, magic wands can be used against you. Robert Cialdini spent years studying the dark arts of social influence: He routinely answered ads recruiting people to work as door-to-door salesmen and telemarketers, and went through their training programs to learn their techniques. He then wrote a manual[23] for those of us who want to resist the tricks of "compliance professionals."

Cialdini describes six principles that salespeople use against us, but the most basic of all is reciprocity. People who want something from us try to give us something first, and we all have piles of address stickers and free postcards from charities that gave them to us out of the goodness of their marketing consultants' hearts. The Hare Krishnas perfected the technique: They pressed flowers or cheap copies of the *Bhagavad Gita* into the hands of unsuspecting pedestrians, and only then asked for a donation. When Cialdini studied the Krishnas at O'Hare Airport in Chicago, he noticed that they routinely went around the garbage pails to collect and recycle the flowers that they knew would be thrown away. Few people wanted the flowers, but in the early days of the technique, most were unable just to accept them and walk on without giving something in return. The Krishnas grew wealthy by exploiting people's reciprocity reflexes—until everyone learned about the Krishnas and found ways to avoid taking the "gift" in the first place.

But legions of others are still after you. Supermarkets and Amway dealers give out free samples to boost sales. Waiters and waitresses put a mint on the check tray, a technique that has been shown to boost tips.[24] Including a five-dollar "gift check" along with a survey sent in the mail increases people's willingness to complete the survey, even more than does promising to send them fifty dollars for completing the survey.[25] If you get something for nothing, part of you may be pleased, but part of you (part of the elephant—automatic processes) moves your hand to your wallet to give something back.

Reciprocity works just as well for bargaining. Cialdini was once asked by a boy scout to buy tickets to a movie he didn't want to see. When Cialdini said no, the scout asked him to buy some less expensive chocolate bars instead. Cialdini found himself walking away with three chocolate bars that he didn't want. The scout had made a concession, and Cialdini automatically reciprocated by making a concession of his own. But rather than getting mad, Cialdini got data. He conducted his own version of the encounter, asking college students walking on campus whether they would volunteer to chaperone a group of juvenile delinquents to the zoo for a day. Only 17 percent agreed. But in another condition of the study, students were first asked whether they would volunteer to work for two hours a week for two years with juvenile delinquents. All said no, but when the experimenter then asked about the day trip to the zoo, 50 percent said yes.[26] Concession leads to concession. In financial bargaining, too, people who stake out an extreme first position and then move toward the middle end up doing better than those who state a more reasonable first position and then hold fast.[27] And the extreme offer followed by concession doesn't just get you a better price, it gets you a happier partner (or victim): She is more likely to honor the agreement because she feels that she had more influence on the outcome. The very process of give and take creates a feeling of partnership, even in the person being taken.

So the next time a salesman gives you a free gift or consultation, or makes a concession of any sort, duck. Don't let him press your reciprocity button. The best way out, Cialdini advises, is to fight reciprocity with reciprocity. If you can reappraise the salesman's move for what it is—an effort to exploit you—you'll feel entitled to exploit him right back. Accept the gift or concession with a feeling of victory—you are exploiting an exploiter—not mindless obligation.

Reciprocity is not just a way of dealing with boy scouts and obnoxious salespeople; it's for friends and lovers, too. Relationships are exquisitely sensitive to balance in their early stages, and a great way to ruin things is either to give too much (you seem perhaps a bit desperate) or too little (you seem cold and rejecting). Rather, relationships grow best by balanced give and take, especially of gifts, favors, attention, and self-disclosure. The first three are somewhat obvious, but people often don't realize the degree to which the disclosure of personal information is a gambit in the dating game. When someone tells you about past romantic relationships, there is conversational pressure for you to do the same. If this disclosure card is played too early, you might feel ambivalence—your reciprocity reflex makes you prepare your own matching disclosure, but some other part of you resists sharing intimate details with a near-stranger. But when it's played at the right time, the past-relationships-mutual-disclosure conversation can be a memorable turning point on the road to love.

Reciprocity is an all-purpose relationship tonic. Used properly, it strengthens, lengthens, and rejuvenates social ties. It works so well in part because the elephant is a natural mimic. For example, when we interact with someone we like, we have a slight tendency to copy their every move, automatically and unconsciously.[28] If the other person taps her foot, you are more likely to tap yours. If she touches her face, you are more likely to touch yours. But it's not just that we mimic those we like; we like those who mimic us. People who are subtly mimicked are then more helpful and agreeable toward their mimicker, and even toward others.[29] Waitresses who mimic their customers get larger tips.[30]

Mimicry is a kind of social glue, a way of saying "We are one." The unifying pleasures of mimicry are particularly clear in synchronized activities, such as line dances, group cheers, and some religious rituals, in which people try to do the same thing at the same time. A theme of the rest of this book is that humans are partially hive creatures, like bees, yet in the modern world we spend nearly all our time outside of the hive. Reciprocity, like love, reconnects us with others.

# 4

# The Faults of Others

> *Why do you see the speck in your neighbor's eye, but do not notice the log in your own eye? . . . You hypocrite, first take the log out of your own eye, and then you will see clearly to take the speck out of your neighbor's eye.*
>
> —MATTHEW 7:3–5

> *It is easy to see the faults of others, but difficult to see one's own faults. One shows the faults of others like chaff winnowed in the wind, but one conceals one's own faults as a cunning gambler conceals his dice.*
>
> —BUDDHA[1]

IT'S FUN TO LAUGH at a hypocrite, and recent years have given Americans a great deal to laugh at. Take the conservative radio show host Rush Limbaugh, who once said, in response to the criticism that the United States prosecutes a disproportionate number of black men for drug crimes, that white drug users should be seized and "sent up the river," too. In 2003, he was forced to eat his words when Florida officials discovered his illegal purchase of massive quantities of Oxycontin, a painkiller also known as "hillbilly heroin." Another case occurred in my home state of Virginia. Congressman Ed Schrock was an outspoken opponent of gay rights, gay marriage, and of gays serving in the military. Speaking of the horrors of such coservice, he said, "I mean, they are in the showers with you, they are in the dining hall with you."[2] In August 2004, audio tapes were made public of the messages that Schrock, a married man, had left on Megamates, an interactive phone sex line. Schrock described the anatomical features of the kind of man he was seeking, along with the acts he was interested in performing.

There is a special pleasure in the irony of a moralist brought down for the very moral failings he has condemned. It's the pleasure of a well-told joke. Some jokes are funny as one-liners, but most require three verses: three guys, say, who walk into a bar one at a time, or a priest, a minister, and a rabbi in a lifeboat. The first two set the pattern, and the third violates it. With hypocrisy, the hypocrite's preaching is the setup, the hypocritical action is the punch line. Scandal is great entertainment because it allows people to feel contempt, a moral emotion that gives feelings of moral superiority while asking nothing in return. With contempt you don't need to right the wrong (as with anger) or flee the scene (as with fear or disgust). And best of all, contempt is made to share. Stories about the moral failings of others are among the most common kinds of gossip,[3] they are a staple of talk radio, and they offer a ready way for people to show that they share a common moral orientation. Tell an acquaintance a cynical story that ends with both of you smirking and shaking your heads and voila, you've got a bond.

Well, stop smirking. One of the most universal pieces of advice from across cultures and eras is that we are all hypocrites, and in our condemnation of others' hypocrisy we only compound our own. Social psychologists have recently isolated the mechanisms that make us blind to the logs in our own eyes. The moral implications of these findings are disturbing; indeed, they challenge our greatest moral certainties. But the implications can be liberating, too, freeing you from destructive moralism and divisive self-righteousness.

## KEEPING UP APPEARANCES

Research on the evolution of altruism and cooperation has relied heavily on studies in which several people (or people simulated on a computer) play a game. On each round of play, one person interacts with one other player and can choose to be cooperative (thereby expanding the pie they then share) or greedy (each grabbing as much as possible for himself). After many rounds of play, you count up the number of points each player accumulated and see which strategy was most profitable in the long run. In these games, which are intended to be simple models of the game of life, no strategy ever beats tit for tat.[4] In the long run and across a variety of environments, it pays to cooperate while remaining vigilant to the danger of being cheated. But those simple games are in some ways simple minded. Players face a binary choice at each point: They can cooperate or defect. Each player then reacts to what the other player did in the previous round. In real life, however, you don't react to what someone did; you react only to what you *think* she did, and the gap between action and perception is bridged by the art of impression management. If life itself is but what you deem it, then why not focus your efforts on persuading others to *believe* that you are a virtuous and trustworthy cooperator? Thus Niccolo Machiavelli, whose name has become synonymous with the cunning and amoral use of power, wrote five hundred years ago that "the great majority of mankind are satisfied with appearances, as though they were realities, and are often more influenced by the things that seem than by those that are."[5] Natural selection, like politics, works by the principle of survival of the fittest, and several researchers have argued that human beings evolved to play the game of life in a Machiavellian way.[6] The Machiavellian version of tit for tat, for example, is to do all you can to cultivate the *reputation* of a trustworthy yet vigilant partner, whatever the reality may be.

The simplest way to cultivate a reputation for being fair is to really be fair, but life and psychology experiments sometimes force us to choose between appearance and reality. Dan Batson at the University of Kansas devised a clever way to make people choose, and his findings are not pretty. He brought students into his lab one at a time to take part in what they thought was a study of how unequal rewards affect teamwork.[7] The procedure was explained: One member of each team of two will be rewarded for correct responses to questions with a raffle ticket that could win a valuable prize. The other member will receive nothing. Subjects were also told that an additional part of the experiment concerned the effects of control: You, the subject, will decide which of you is rewarded, which of you is not. Your partner is already here, in another room, and the two of you will not meet. Your partner will be told that the decision was made by chance. You can make the decision in any way you like. Oh, and here is a coin: Most people in this study seem to think that flipping the coin is the fairest way to make the decision.

Subjects were then left alone to choose. About half of them used the coin. Batson knows this because the coin was wrapped in a plastic bag, and half the bags were ripped open. Of those who did not flip the coin, 90 percent chose the positive task for themselves. For those who did flip the coin, the laws of probability were suspended and 90 percent of them chose the positive task for themselves. Batson had given all the subjects a variety of questionnaires about morality weeks earlier (the subjects were students in psychology classes), so he was able to check how various measures of moral personality predicted behavior. His finding: People who reported being most concerned about caring for others and about issues of social responsibility were more likely to open the bag, but they were not more likely to give the other person the positive task. In other words, people who think they are particularly moral are in fact more likely to "do the right thing" and flip the coin, but when the coin flip comes out against them, they find a way to ignore it and follow their own self-interest. Batson called this tendency to value the appearance of morality over the reality "moral hypocrisy."

Batson's subjects who flipped the coin reported (on a questionnaire) that they had made the decision in an ethical way. After his first study, Batson wondered whether perhaps people tricked themselves by not stating clearly what heads or tails would mean ("Let's see, heads, that means, um, oh yeah, I get the good one."). But when he labeled the two sides of the coin to erase ambiguity, it made no difference. Placing a large mirror in the room, right in front of the subject, and at the same time stressing the importance of fairness in the instructions, was the only manipulation that had an effect. When people were forced to think about fairness and could see themselves cheating, they stopped doing it. As Jesus and Buddha said in the opening epigraphs of this chapter, it is easy to spot a cheater when our eyes are looking outward, but hard when looking inward. Folk wisdom from around the world concurs:

Though you see the seven defects of others, we do not see our own ten defects. (Japanese proverb)[8]

A he-goat doesn't realize that he smells. (Nigerian proverb)[9]

Proving that people are selfish, or that they'll sometimes cheat when they know they won't be caught, seems like a good way to get an article into the *Journal of Incredibly Obvious Results.* What's not so obvious is that, in nearly all these studies, people don't think they are doing anything wrong. It's the same in real life. From the person who cuts you off on the highway all the way to the Nazis who ran the concentration camps, most people think they are good people and that their actions are motivated by good reasons. Machiavellian tit for tat requires devotion to appearances, including protestations of one's virtue even when one chooses vice. And such protestations are most effective when the person making them really believes them. As Robert Wright put it in his masterful book *The Moral Animal,* "Human beings are a species splendid in their array of moral equipment, tragic in their propensity to misuse it, and pathetic in their constitutional ignorance of the misuse."[10]

If Wright is correct about our "constitutional ignorance" of our hypocrisy, then the sages' admonition to stop smirking may be no more effective than telling a depressed person to snap out of it. You can't change your mental filters by willpower alone; you have to engage in activities such as meditation or cognitive therapy that train the elephant. But at least a depressed person will usually admit she's depressed. Curing hypocrisy is much harder because part of the problem is that we don't believe there's a problem. We are well-armed for battle in a Machiavellian world of reputation manipulation, and one of our most important weapons is the delusion that we are noncombatants. How do we get away with it?

## FIND YOUR INNER LAWYER

Remember Julie and Mark, the sister and brother who had sex back in chapter 1? Most people condemned their actions even in the absence of harm,

and then made up reasons, sometimes bad ones, to justify their condemnation. In my studies of moral judgment, I have found that people are skilled at
finding reasons to support their gut feelings: The rider acts like a lawyer
whom the elephant has hired to represent it in the court of public opinion.

One of the reasons people are often contemptuous of lawyers is that
they fight for a client's interests, not for the truth. To be a good lawyer, it
often helps to be a good liar. Although many lawyers won't tell a direct lie,
most will do what they can to hide inconvenient facts while weaving a
plausible alternative story for the judge and jury, a story that they sometimes know is not true. Our inner lawyer works in the same way, but,
somehow, we actually believe the stories he makes up. To understand his
ways we must catch him in action; we must observe him carrying out low-
pressure as well as high-pressure assignments.

People sometimes call their lawyers to ask whether a particular course of
action is permissible. No pressure, just tell me whether I can do this. The
lawyer looks into the relevant laws and procedures and calls back with a ver-
dict: Yes, there is a legal or regulatory precedent for that; or perhaps no, as
your lawyer I would advise against such a course. A good lawyer might look
at all sides of a question, think about all possible ramifications, and recom-
mend alternative courses of action, but such thoroughness depends in part
on his client—does she really want advice or does she just want to be given
a red or a green light for her plan?

Studies of everyday reasoning show that the elephant is not an inquisitive
client. When people are given difficult questions to think about—for example, whether the minimum wage should be raised—they generally lean
one way or the other right away, and then put a call in to reasoning to see
whether support for that position is forthcoming. For example, a person
whose first instinct is that the minimum wage should be raised looks around
for supporting evidence. If she thinks of her Aunt Flo who is working for
the minimum wage and can't support her family on it then yes, that means
the minimum wage should be raised. All done. Deanna Kuhn,[11] a cognitive
psychologist who has studied such everyday reasoning, found that most
people readily offered "pseudoevidence" like the anecdote about Aunt Flo.
Most people gave no real evidence for their positions, and most made no effort to look for evidence opposing their initial positions. David Perkins,[12] a

Harvard psychologist who has devoted his career to improving reasoning, found the same thing. He says that thinking generally uses the "makes-sense" stopping rule. We take a position, look for evidence that supports it, and if we find some evidence—enough so that our position "makes sense"—we stop thinking. But at least in a low-pressure situation such as this, if someone *else* brings up reasons and evidence on the other side, people can be induced to change their minds; they just don't make an effort to do such thinking for themselves.

Now let's crank up the pressure. The client has been caught cheating on her taxes. She calls her lawyer. She doesn't confess and ask, "Was that OK?" She says, "Do something." The lawyer bolts into action, assesses the damaging evidence, researches precedents and loopholes, and figures out how some personal expenses might be plausibly justified as business expenses. The lawyer has been given an order: Use all your powers to defend me. Studies of "motivated reasoning"[13] show that people who are motivated to reach a particular conclusion are even worse reasoners than those in Kuhn's and Perkins's studies, but the mechanism is basically the same: a one-sided search for supporting evidence only. People who are told that they have performed poorly on a test of social intelligence think extra hard to find reasons to discount the test; people who are asked to read a study showing that one of their habits—such as drinking coffee—is unhealthy think extra hard to find flaws in the study, flaws that people who don't drink coffee don't notice. Over and over again, studies show that people set out on a cognitive mission to bring back reasons to support their preferred belief or action. And because we are usually successful in this mission, we end up with the illusion of objectivity. We really believe that our position is rationally and objectively justified.

Ben Franklin, as usual, was wise to our tricks. But he showed unusual insight in catching himself in the act. Though he had been a vegetarian on principle, on one long sea crossing the men were grilling fish, and his mouth started watering:

I balanc'd some time between principle and inclination, till I recollectd that, when the fish were opened, I saw smaller fish taken out of their stomachs; then thought I, "if you eat one another, I don't see why we mayn't eat you." So I din'd upon cod very heartily, and continued to eat with other people, returning only now and then occasionally to a vegetable diet.[14]

Franklin concluded: "So convenient a thing is it to be a reasonable creature, since it enables one to find or make a reason for every thing one has a mind to do."

## THE ROSE-COLORED MIRROR

I don't want to blame everything on the lawyer. The lawyer is, after all, the rider—your conscious, reasoning self; and he is taking orders from the elephant—your automatic and unconscious self. The two are in cahoots to win at the game of life by playing Machiavellian tit for tat, and both are in denial about it.

To win at this game you must present your best possible self to others. You must appear virtuous, whether or not you are, and you must gain the benefits of cooperation whether or not you deserve them. But everyone else is playing the same game, so you must also play defense—you must be wary of others' self-presentations, and of their efforts to claim more for themselves than they deserve. Social life is therefore always a game of social comparison. We must compare ourselves to other people, and our actions to their actions, and we must somehow spin those comparisons in our favor. (In depression, part of the illness is that spin goes the other way, as described by Aaron Beck's cognitive triad: I'm bad, the world is terrible, and my future is bleak.) You can spin a comparison either by inflating your own claims or by disparaging the claims of others. You might expect, given what I've said so far, that we do both, but the consistent finding of psychological research is that we are fairly accurate in our perceptions of others. It's our self-perceptions that are distorted because we look at ourselves in a rose-colored mirror.

In Garrison Keillor's mythical town of Lake Wobegon, all the women are strong, all the men good looking, and all the children above average. But if the Wobegonians were real people, they would go further: Most of them would believe they were stronger, better looking, or smarter than the average Wobegonian. When Americans and Europeans are asked to rate themselves on virtues, skills, or other desirable traits (including intelligence, driving ability, sexual skills, and ethics), a large majority say they are above average.[15] (This effect is weaker in East Asian countries, and may not exist in Japan.)[16]

In a brilliant series of experiments,[17] Nick Epley and David Dunning figured out how we do it. They asked students at Cornell University to predict how many flowers they would buy in an upcoming charity event and how many the average Cornell student would buy. Then they looked at actual behavior. People had greatly overestimated their own virtue, but were pretty close on their guesses about others. In a second study, Epley and Dunning asked people to predict what they would do in a game that could be played for money either selfishly or cooperatively. Same findings: Eighty-four percent predicted that they'd cooperate, but the subjects expected (on average) that only 64 percent of others would cooperate. When they ran the real game, 61 percent cooperated. In a third study, Epley and Dunning paid people five dollars for participating in an experiment and then asked them to predict how much of the money they and others would donate, hypothetically, had they been given a particular charitable appeal after the study. People said (on average) they'd donate $2.44, and others would donate only $1.83. But when the study was rerun with a real request to give money, the average gift was $1.53.

In their cleverest study, the researchers described the details of the third study to a new group of subjects and asked them to predict how much money they would donate if they had been in the "real" condition, and how much money other Cornell students would donate. Once again, subjects predicted they'd be much more generous than others. But then subjects saw the actual amounts of money donated by real subjects from the third study, revealed to them one at a time (and averaging $1.53). After being given this new information, subjects were given a chance to revise their estimates, and they did. They lowered their estimates of what others would give, but they did not change their estimates of what they themselves would give. In other words, subjects used base rate information properly to revise their predictions of *others*, but they refused to apply it to their rosy self-assessments. We judge others by their behavior, but we think we have special information about ourselves—we know what we are "really like" inside, so we can easily find ways to explain away our selfish acts and cling to the illusion that we are better than others.

Ambiguity abets the illusion. For many traits, such as leadership, there are so many ways to define it that one is free to pick the criterion that will most flatter oneself. If I'm confident, I can define leadership as confidence. If I think I'm high on people skills, I can define leadership as the ability to understand and influence people. When comparing ourselves to others, the general process is this: Frame the question (unconsciously, automatically) so that the trait in question is related to a self-perceived strength, then go out and look for evidence that you have the strength. Once you find a piece of evidence, once you have a "makes-sense" story, you are done. You can stop thinking, and revel in your self-esteem. It's no wonder, then, that in a study of 1 million American high school students, 70 percent thought they were above average on leadership ability, but only 2 percent thought they were below average. Everyone can find *some* skill that might be construed as related to leadership, and then find *some* piece of evidence that one has that skill.[18] (College professors are less wise than high school students in this respect—94 percent of us think we do above-average work.)[19] But when there is little room for ambiguity—how tall are you? how good are you at juggling?—people tend to be much more modest.

If the only effect of these rampant esteem-inflating biases was to make people feel good about themselves, they would not be a problem. In fact, evidence shows that people who hold pervasive positive illusions about themselves, their abilities, and their future prospects are mentally healthier, happier, and better liked than people who lack such illusions.[20] But such biases can make people feel that they deserve more than they do, thereby setting the stage for endless disputes with other people who feel equally over-entitled.

I fought endlessly with my first-year college roommates. I had provided much of our furniture, including the highly valued refrigerator, and I did most of the work keeping our common space clean. After a while, I got tired of doing more than my share; I stopped working so hard and let the space become messy so that someone else would pick up the slack. Nobody did. But they did pick up my resentment, and it united them in their dislike of me. The next year, when we no longer lived together, we became close friends.

When my father drove me and my refrigerator up to college that first year, he told me that the most important things I was going to learn I would not learn in the classroom, and he was right. It took many more years of living with roommates, but I finally realized what a fool I had made of myself that first year. Of course I thought I did more than my share. Although I was aware of every little thing I did for the group, I was aware of only a portion of everyone else's contributions. And even if I had been correct in my accounting, I was self-righteous in setting up the accounting categories. I picked the things I cared about—such as keeping the refrigerator clean—and then gave myself an A-plus in that category. As with other kinds of social comparison, ambiguity allows us to set up the comparison in ways that favor ourselves, and then to seek evidence that shows we are excellent cooperators. Studies of such "unconscious overclaiming" show that when husbands and wives estimate the percentage of housework each does, their estimates total more than 120 percent.[21] When MBA students in a work group make estimates of their contributions to the team, the estimates total 139 percent.[22] Whenever people form cooperative groups, which are usually of mutual benefit, self-serving biases threaten to fill group members with mutual resentment.

## I'M RIGHT; YOU'RE BIASED

If spouses, colleagues, and roommates so easily descend into resentment, things get worse when people who lack affection or shared goals have to negotiate. Vast societal resources are expended on litigation, labor strikes, divorce disputes, and violence after failed peace talks because the same self-serving biases are at work fomenting hypocritical indignation. In these high-pressure situations, the lawyers (real and metaphorical) are working round the clock to spin and distort the case in their clients' favor. George Loewenstein[23] and his colleagues at Carnegie Mellon found a way to study the process by giving pairs of research subjects a real legal case to read (about a motorcycle accident in Texas), assigning one subject to play the defendant and one the plaintiff, and then giving them real money to nego-tiate with. Each pair was told to reach a fair agreement and warned that, if they failed to agree, a settlement would be imposed and "court costs" de-ducted from the pool of money, leaving both players worse off. When both players knew which role each was to play from the start, each read the case materials differently, made different guesses about what settlement the judge in the real case had imposed, and argued in a biased way. More than a quarter of all pairs failed to reach an agreement. However, when the play-ers didn't know which role they were to play until after they had read all the materials, they became much more reasonable, and only 6 percent of pairs failed to settle.

Recognizing that hiding negotiators' identities from them until the last minute is not an option in the real world, Loewenstein set out to find other ways to "de-bias" negotiators. He tried having subjects read a short essay about the kinds of self-serving biases that affect people in their situation to see whether subjects could correct for the biases. No dice. Although the subjects used the information to predict their opponent's behavior more accurately, they did not change their own biases at all. As Epley and Dun-ning had found, people really are open to information that will predict the behavior of others, but they refuse to adjust their self-assessments. In an-other study, Loewenstein followed the advice often given by marriage ther-apists to have each subject first write an essay arguing the other person's case as convincingly as possible. Even worse than no dice. The manipula-tion backfired, perhaps because thinking about your opponent's arguments automatically triggers additional thinking on your own part as you prepare to refute them.

One manipulation did work. When subjects read the essay about self-serving biases and were then asked to write an essay about weaknesses in *their own* case, their previous righteousness was shaken. Subjects in this study were just as fair-minded as those who learned their identities at the last minute. But before you get too optimistic about this technique for re-ducing hypocrisy, you should realize that Loewenstein was asking subjects to find weaknesses in their *cases*—in the positions they were arguing for—not in their *characters*. When you try to persuade people to look at their own personal picture of Dorian Gray, they put up a much bigger fight. Emily Pronin at Princeton and Lee Ross at Stanford have tried to help people overcome their self-serving biases by teaching them about biases and then asking, "OK, now that you know about these biases, do you want to change what you just said about yourself?" Across many studies, the results were the same:[24] People were quite happy to learn about the various forms of self-serving bias and then apply their newfound knowledge to predict others' responses. But their self-ratings were unaffected. Even when you grab people by the lapels, shake them, and say, "Listen to me! Most people have an inflated view of themselves. Be realistic!" they refuse, muttering to themselves, "Well, other people may be biased, but I *really am* above average on leadership."

Pronin and Ross trace this resistance to a phenomenon they call "naive realism": Each of us thinks we see the world directly, as it really is. We further believe that the facts as we see them are there for all to see, therefore others should agree with us. If they don't agree, it follows either that they have not yet been exposed to the relevant facts or else that they are blinded by their interests and ideologies. People acknowledge that their own backgrounds have shaped their views, but such experiences are invariably seen as deepening one's insights; for example, being a doctor gives a person special insight into the problems of the health-care industry. But the background of other people is used to explain their biases and covert motivations; for example, doctors think that lawyers disagree with them about tort reform not because they work with the victims of malpractice (and therefore have their own special insights) but because their self-interest biases their thinking. It just seems plain as day, to the naive realist, that everyone is influenced by ideology and self-interest. Except for me. I see things as they are.

If I could nominate one candidate for "biggest obstacle to world peace and social harmony," it would be naive realism because it is so easily ratcheted up from the individual to the group level: My group is right because we see things as they are. Those who disagree are obviously biased by their religion, their ideology, or their self-interest. Naive realism gives us a world full of good and evil, and this brings us to the most disturbing implication of the sages' advice about hypocrisy: Good and evil do not exist outside of our beliefs about them.

## SATAN SATISFIES

One day in 1998 I received a handwritten letter from a woman in my town whom I did not know. The woman wrote about how crime, drugs, and teen pregnancy were all spiraling out of control. Society was going downhill as Satan spread his wings. The woman invited me to come to her church and find spiritual shelter. As I read her letter, I had to agree with her that Satan had spread his wings, but only to fly away and leave us in peace. The late 1990s was a golden age. The cold war was over, democracy and human rights were spreading, South Africa had vanquished apartheid, Israelis and Palestinians were reaping the fruits of the Oslo accords, and there were encouraging signs from North Korea. Here in the United States, crime and unemployment had plummeted, the stock market was climbing ever higher, and the ensuing prosperity was promising to erase the national debt. Even cockroaches were disappearing from our cities because of widespread use of the roach poison Combat. So what on earth was she talking about?

When the moral history of the 1990s is written, it might be titled *Desperately Seeking Satan*. With peace and harmony ascendant, Americans seemed to be searching for substitute villains. We tried drug dealers (but then the crack epidemic waned) and child abductors (who are usually one of the parents). The cultural right vilified homosexuals; the left vilified racists and homophobes. As I thought about these various villains, including the older villains of communism and Satan himself, I realized that most of them share three properties: They are invisible (you can't identify the evil one from appearance alone); their evil spreads by contagion, making it vital to protect impressionable young people from infection (for example from communist ideas, homosexual teachers, or stereotypes on television); and the villains can be defeated only if we all pull together as a team. It became clear to me that people want to believe they are on a mission from God, or that they are fighting for some more secular good (animals, fetuses, women's rights), and you can't have much of a mission without good allies and a good enemy.

The problem of evil has bedeviled many religions since their birth. If God is all good and all powerful, either he allows evil to flourish (which means he is not all good), or else he struggles against evil (which means he is not all powerful). Religions have generally chosen one of three resolutions of this paradox.[25] One solution is straight dualism: There exists a good force and an evil force, they are equal and opposite, and they fight eternally. Human beings are part of the battleground. We were created part good, part evil, and we must choose which side we will be on. This view is clearest in religions emanating from Persia and Babylonia, such as Zoroastrianism, and the view influenced Christianity as a long-lived doctrine called Manichaeism. A second resolution is straight monism: There is one God; he created the world as it needs to be, and evil is an illusion, a view that dominated religions that developed in India. These religions hold that the entire world—or, at least, its emotional grip upon us—is an illusion, and that enlightenment consists of breaking out of the illusion. The third approach, taken by Christianity, blends monism and dualism in a way that ultimately reconciles the goodness and power of God with the existence of Satan. This argument is so complicated that I cannot understand it. Nor, apparently, can many Christians who, judging by what I hear on gospel radio stations in Virginia, seem to hold a straight Manichaean world view, according to which God and Satan are fighting an eternal war. In fact, despite the diversity of theological arguments made in different religions, concrete representations of Satan, demons, and other evil entities are surprisingly similar across continents and eras.[26]

From a psychological perspective, Manichaeism makes perfect sense. "Our life is the creation of our mind," as Buddha said, and our minds evolved to play Machiavellian tit for tat. We all commit selfish and shortsighted acts, but our inner lawyer ensures that we do not blame ourselves or our allies for them. We are thus convinced of our own virtue, but quick to see bias, greed, and duplicity in others. We are often correct about others' motives, but as any conflict escalates we begin to exaggerate grossly, to weave a story in which pure virtue (our side) is in a battle with pure vice (theirs).

## THE MYTH OF PURE EVIL

In the days after receiving that letter, I thought a lot about the need for evil. I decided to write an article on this need and use the tools of modern psychology to understand evil in a new way. But as soon as I started my research, I found out I was too late. By one year. A three-thousand-year-old question had been given a complete and compelling psychological explanation the previous year by Roy Baumeister, one of today's most creative social psychologists. In *Evil: Inside Human Cruelty and Aggression*,[27] Baumeister examined evil from the perspective of both victim and perpetrator. When taking the perpetrator's perspective, he found that people who do things we see as evil, from spousal abuse all the way to genocide, rarely think they are doing anything wrong. They almost always see themselves as responding to attacks and provocations in ways that are justified. They often think that they themselves are victims. But, of course, you can see right through this tactic; you are good at understanding the biases that others use to protect their self-esteem. The disturbing part is that Baumeister shows us our own distortions as victims, and as righteous advocates of victims. Almost everywhere Baumeister looked in the research literature, he found that victims often shared some of the blame. Most murders result from an escalating cycle of provocation and retaliation; often, the corpse could just as easily have been the murderer. In half of all domestic disputes, both sides used violence.[28] Baumeister points out that, even in instances of obvious police brutality, such as the infamous videotaped beating of Rodney King in Los Angeles in 1991, there is usually much more to the story than is shown on the news. (News programs gain viewers by satisfying people's need to believe that evil stalks the land.)

Baumeister is an extraordinary social psychologist, in part because in his search for truth he is unconcerned about political correctness. Sometimes evil falls out of a clear blue sky onto the head of an innocent victim, but most cases are much more complicated, and Baumeister is willing to violate the taboo against "blaming the victim" in order to understand what really happened. People usually have reasons for committing violence, and those reasons usually involve retaliation for a perceived injustice, or self-defense. This does not mean that both sides are equally to blame: Perpetrators often grossly overreact and misinterpret (using self-serving biases). But Baumeister's point is that we have a deep need to understand violence and cruelty through what he calls "the myth of pure evil." Of this myth's many parts, the most important are that evildoers are pure in their evil motives (they have no motives for their actions beyond sadism and greed);

victims are pure in their victimhood (they did nothing to bring about their victimization); and evil comes from outside and is associated with a group or force that attacks our group. Furthermore, anyone who questions the application of the myth, who dares muddy the waters of moral certainty, is in league with evil.

The myth of pure evil is the ultimate self-serving bias, the ultimate form of naive realism. And it is the ultimate cause of most long-running cycles of violence because both sides use it to lock themselves into a Manichaean struggle. When George W. Bush said that the 9/11 terrorists did what they did because they "hate our freedom," he showed a stunning lack of psychological insight. Neither the 9/11 hijackers nor Osama Bin Laden were particularly upset because American women can drive, vote, and wear bikinis. Rather, many Islamic extremists want to kill Americans because they are using the Myth of Pure Evil to interpret Arab history and current events. They see America as the Great Satan, the current villain in a long pageant of Western humiliation of Arab nations and peoples. They did what they did as a reaction to America's actions and impact in the Middle East, as they see it through the distortions of the Myth of Pure Evil. However horrifying it is for terrorists to lump all civilians into the category of "enemy" and then kill them indiscriminately, such actions at least make psychological sense, whereas killing because of a hatred for freedom does not.

In another unsettling conclusion, Baumeister found that violence and cruelty have four main causes. The first two are obvious attributes of evil: greed/ambition (violence for direct personal gain, as in robbery) and sadism (pleasure in hurting people). But greed/ambition explains only a small portion of violence, and sadism explains almost none. Outside of children's cartoons and horror films, people almost never hurt others for the sheer joy of hurting someone. The two biggest causes of evil are two that we think are good, and that we try to encourage in our children: high self-esteem and moral idealism. Having high self-esteem doesn't directly cause violence, but when someone's high esteem is unrealistic or narcissistic, it is easily threatened by reality; in reaction to those threats, people—particularly young men—often lash out violently.[29] Baumeister questions the usefulness of programs that try raise children's self-esteem directly instead of by teaching them skills they can be proud of. Such direct enhancement can potentially foster unstable narcissism.

Threatened self-esteem accounts for a large portion of violence at the individual level, but to really get a mass atrocity going you need idealism—the belief that your violence is a means to a moral end. The major atrocities of the twentieth century were carried out largely either by men who thought they were creating a utopia or else by men who believed they were defending their homeland or tribe from attack.[30] Idealism easily becomes dangerous because it brings with it, almost inevitably, the belief that the ends justify the means. If you are fighting for good or for God, what matters is the outcome, not the path. People have little respect for rules; we respect the moral principles that underlie most rules. But when a moral mission and legal rules are incompatible, we usually care more about the mission. The psychologist Linda Skitka[31] finds that when people have strong moral feelings about a controversial issue—when they have a "moral mandate"—they care much less about procedural fairness in court cases. They want the "good guys" freed by any means, and the "bad guys" convicted by any means. It is thus not surprising that the administration of George W. Bush consistently argues that extra-judicial killings, indefinite imprisonment without trial, and harsh physical treatment of prisoners are legal and proper steps in fighting the Manichaean "war on terror."

## FINDING THE GREAT WAY

In philosophy classes, I often came across the idea that the world is an illusion. I never really knew what that meant, although it sounded deep. But after two decades studying moral psychology, I think I finally get it. The anthropologist Clifford Geertz wrote that "man is an animal suspended in webs of significance that he himself has spun."[32] That is, the world we live in is not really one made of rocks, trees, and physical objects; it is a world of insults, opportunities, status symbols, betrayals, saints, and sinners. All of these are human creations which, though real in their own way, are not real in the way that rocks and trees are real. These human creations are like fairies in J. M. Barrie's *Peter Pan:* They exist only if you believe in them. They are the Matrix (from the movie of that name); they are a consensual hallucination.

The inner lawyer, the rose-colored mirror, naive realism, and the myth of pure evil—these mechanisms all conspire to weave for us a web of significance upon which angels and demons fight it out. Our ever-judging minds then give us constant flashes of approval and disapproval, along with the certainty that we are on the side of the angels. From this vantage point it all seems so silly, all this moralism, righteousness, and hypocrisy. It's beyond silly; it is tragic, for it suggests that human beings will never achieve a state of lasting peace and harmony. So what can you do about it?

The first step is to see it as a game and stop taking it so seriously. The great lesson that comes out of ancient India is that life as we experience it is a game called "samsara." It is a game in which each person plays out his "dharma," his role or part in a giant play. In the game of samsara, good things happen to you, and you are happy. Then bad things happen, and you are sad or angry. And so it goes, until you die. Then you are reborn back into it, and it repeats. The message of the *Bhagavad Gita* (a central text of Hinduism) is that you can't quit the game entirely; you have a role to play in the functioning of the universe, and you must play that role. But you should do it in the right way, without being attached to the "fruits" or outcomes of your action. The god Krishna says:

> I love the man who hates not nor exults, who mourns not nor desires . . . who is the same to friend and foe, [the same] whether he be respected or despised, the same in heat and cold, in pleasure and in pain, who has put away attachment and remains unmoved by praise or blame . . . contented with whatever comes his way.[33]

Buddha went a step further. He, too, counseled indifference to the ups and downs of life, but he urged that we quit the game entirely. Buddhism is a set of practices for escaping samsara and the endless cycle of rebirth. Though divided on whether to retreat from the world or engage with it, Buddhists all agree on the importance of training the mind to stop its incessant judging. Sen-ts'an, an early Chinese Zen master, urged nonjudgmentalism as a prerequisite to following "the perfect way" in this poem from the eighth century CE:

> The Perfect Way is only difficult for those who pick and
>     choose;
> Do not like, do not dislike; all will then be clear.
> Make a hairbreadth difference, and Heaven and Earth are
>     set apart;
> If you want the truth to stand clear before you, never be for
>     or against.
> The struggle between "for" and "against" is the mind's worst
>     disease.[34]

Judgmentalism is indeed a disease of the mind: it leads to anger, torment, and conflict. But it is also the mind's normal condition—the elephant is always evaluating, always saying "Like it" or "Don't like it." So how can you change your automatic reactions? You know by now that you can't simply resolve to stop judging others or to stop being a hypocrite. But, as Buddha taught, the rider can gradually learn to tame the elephant, and meditation is one way to do so. Meditation has been shown to make people calmer, less reactive to the ups and downs and petty provocations of life.[35] Meditation is the Eastern way of training yourself to take things philosophically.

Cognitive therapy works, too. In *Feeling Good*,[36] a popular guide to cognitive therapy, David Burns has written a chapter on cognitive therapy for anger. He advises using many of the same techniques that Aaron Beck used for depression: Write down your thoughts, learn to recognize the distortions in your thoughts, and then think of a more appropriate thought. Burns focuses on the *should* statements we carry around—ideas about how the world *should* work, and about how people *should* treat us. Violations of these *should* statements are the major causes of anger and resentment. Burns also advises empathy: In a conflict, look at the world from your opponent's point of view, and you'll see that she is not entirely crazy.

Although I agree with Burns's general approach, the material I have reviewed in this chapter suggests that, once anger comes into play, people find it extremely difficult to empathize with and understand another perspective. A better place to start is, as Jesus advised, with yourself and the log in your own eye. (Batson and Loewenstein both found that debiasing occurred only when subjects were forced to look at themselves.) And you will see the log only if you set out on a deliberate and effortful quest to look for it. Try this now: Think of a recent interpersonal conflict with someone you care about and then find one way in which your behavior was not exemplary. Maybe you did something insensitive (even if you had a right to do it), or hurtful (even if you meant well), or inconsistent with your principles (even though you can readily justify it). When you first catch sight of a fault in yourself, you'll likely hear frantic arguments from your inner lawyer excusing you and blaming others, but try not to listen. You are on a mission to find at least one thing that you did wrong. When you extract a splinter it hurts, briefly, but then you feel relief, even pleasure. When you find a fault in yourself it will hurt, briefly, but if you keep going and acknowledge the fault, you are likely to be rewarded with a flash of pleasure that is mixed, oddly, with a hint of pride. It is the pleasure of taking responsibility for your own behavior. It is the feeling of honor.

Finding fault with yourself is also the key to overcoming the hypocrisy and judgmentalism that damage so many valuable relationships. The instant you see some contribution you made to a conflict, your anger softens—maybe just a bit, but enough that you might be able to acknowledge some merit on the other side. You can still believe you are right and the other person is wrong, but if you can move to believing that you are *mostly* right, and your opponent is *mostly* wrong, you have the basis for an effective and nonhumiliating apology. You can take a small piece of the disagreement and say, "I should not have done X, and I can see why you felt Y." Then, by the power of reciprocity, the other person will likely feel a strong urge to say, "Yes, I was really upset by X. But I guess I shouldn't have done P, so I can see why you felt Q." Reciprocity amplified by self-serving biases drove you apart back when you were matching insults or hostile gestures, but you can turn the process around and use reciprocity to end a conflict and save a relationship.

The human mind may have been shaped by evolutionary processes to play Machiavellian tit for tat, and it seems to come equipped with cognitive processes that predispose us to hypocrisy, self-righteousness, and moralistic conflict. But sometimes, by knowing the mind's structure and strategies, we can step out of the ancient game of social manipulation and enter into a game of our choosing. By seeing the log in your own eye you can become less biased, less moralistic, and therefore less inclined toward argument and conflict. You can begin to follow the perfect way, the path to happiness that leads through acceptance, which is the subject of the next chapter.

# 5

## The Pursuit of Happiness

*Good men, at all times, surrender in truth all attachments.
The holy spend not idle words on things of desire. When
pleasure or pain comes to them, the wise feel above pleasure
and pain.*

— BUDDHA[1]

*Do not seek to have events happen as you want them to, but
instead want them to happen as they do happen, and your
life will go well.*

— EPICTETUS[2]

IF MONEY OR POWER could buy happiness, then the author of the Old Testament book of Ecclesiastes should have been overjoyed. The text attributes itself to a king in Jerusalem, who looks back on his life and his search for happiness and fulfillment. He tried at one point to "make a test of pleasure," by seeking happiness in his riches:

I made great works; I built houses and planted vineyards for myself; I made myself gardens and parks, and planted in them all kinds of fruit trees . . . I also had great possessions of herds and flocks, more than any who had been before me in Jerusalem. I also gathered for myself silver and gold and the treasure of kings and of the provinces; I got singers, both men and women, and delights of the flesh, and many concubines. So I became great and surpassed all who were before me in Jerusalem; also my wisdom remained with me. Whatever my eyes desired I did not keep from them. (ECCLESIASTES 2:4–10)

But in what may be one of the earliest reports of a midlife crisis, the author finds it all pointless:

Then I considered all that my hands had done and the toil I had spent in doing it, and again, all was vanity and a chasing after wind, and there was nothing to be gained under the sun. (ECCLESIASTES 2:11)

The author tells us about many other avenues he pursued—hard work, learning, wine—but nothing brought satisfaction; nothing could banish the feeling that his life had no more intrinsic worth or purpose than that of an animal. From the perspective of Buddha and the Stoic philosopher Epictetus, the author's problem is obvious: his *pursuit* of happiness. Buddhism and Stoicism teach that striving for external goods, or to make the world conform to your wishes, is always a striving after wind. Happiness can only be found within, by breaking attachments to external things and cultivating an attitude of acceptance. (Stoics and Buddhists can have relationships, jobs, and possessions, but, to avoid becoming upset upon losing them, they must not be emotionally attached to them.) This idea is of course an extension of the truth of chapter 2: life itself is but what you deem it, and your mental state determines how you deem things. But recent research in psychology suggests that Buddha and Epictetus may have taken things too far. Some things are worth striving for, and happiness comes in part from outside of yourself, if you know where to look.

## THE PROGRESS PRINCIPLE

The author of Ecclesiastes wasn't just battling the fear of meaninglessness; he was battling the disappointment of success. The pleasure of getting

what you want is often fleeting. You dream about getting a promotion, being accepted into a prestigious school, or finishing a big project. You work
every waking hour, perhaps imagining how happy you'd be if you could just
achieve that goal. Then you succeed, and if you're lucky you get an hour,
maybe a day, of euphoria, particularly if your success was unexpected and
there was a moment in which it was revealed (. . . the envelope, please).
More typically, however, you don't get any euphoria. When success seems
increasingly probable and some final event confirms what you already had
begun to expect, the feeling is more one of relief—the pleasure of closure
and release. In such circumstances, my first thought is seldom "Hooray!
Fantastic!"; it is "Okay, what do I have to do now?"

My underjoyed response to success turns out to be normal. And from an
evolutionary point of view, it's even sensible. Animals get a rush of dopamine,
the pleasure neurotransmitter, whenever they do something that advances
their evolutionary interests and moves them ahead in the game of life. Food
and sex give pleasure, and that pleasure serves as a reinforcer (in behaviorist
terms) that motivates later efforts to find food and sex. For humans, however,
the game is more complex. People win at the game of life by achieving high
status and a good reputation, cultivating friendships, finding the best
mate(s), accumulating resources, and rearing their children to be successful
at the same game. People have many goals and therefore many sources of
pleasure. So you'd think we would receive an enormous and long-lasting shot
of dopamine whenever we succeed at an important goal. But here's the trick
with reinforcement: It works best when it comes seconds—not minutes or
hours—after the behavior. Just try training your dog to fetch by giving him a
big steak ten minutes after each successful retrieval. It can't be done.

The elephant works the same way: *It feels pleasure whenever it takes a step
in the right direction.* The elephant learns whenever pleasure (or pain) follows
immediately after behavior, but it has trouble connecting success on Friday
with actions it took on Monday. Richard Davidson, the psychologist who
brought us affective style and the approach circuits of the front left cortex,
writes about two types of positive affect. The first he calls "pre-goal attainment positive affect," which is the pleasurable feeling you get as you make
progress toward a goal. The second is called "post-goal attainment positive affect," which Davidson says arises once you have achieved something you

want.[3] You experience this latter feeling as contentment, as a short-lived feeling of release when the left prefrontal cortex reduces its activity after a goal has been achieved. In other words, when it comes to goal pursuit, it really is the journey that counts, not the destination. Set for yourself any goal you want. Most of the pleasure will be had along the way, with every step that takes you closer. The final moment of success is often no more thrilling than the relief of taking off a heavy backpack at the end of a long hike. If you went on the hike only to feel that pleasure, you are a fool. Yet people sometimes do just this. They work hard at a task and expect some special euphoria at the end. But when they achieve success and find only moderate and short-lived pleasure, they ask (as the singer Peggy Lee once did): Is that all there is? They devalue their accomplishments as a striving after wind.

We can call this "the progress principle": Pleasure comes more from making progress toward goals than from achieving them. Shakespeare captured it perfectly: "Things won are done; joy's soul lies in the doing."[4]

## THE ADAPTATION PRINCIPLE

If I gave you ten seconds to name the very best and very worst things that could ever happen to you, you might well come up with these: winning a 20-million-dollar lottery jackpot and becoming paralyzed from the neck down. Winning the lottery would bring freedom from so many cares and limitations; it would enable you to pursue your dreams, help others, and live in comfort, so it ought to bring long-lasting happiness rather than one serving of dopamine. Losing the use of your body, on the other hand, would bring more limitations than life in prison. You'd have to give up on nearly all your goals and dreams, forget about sex, and depend on other people for help with eating and bathroom functions. Many people think they would rather be dead than paraplegic. But they are mistaken.

Of course, it's better to win the lottery than to break your neck, but not by as much as you'd think. Because whatever happens, you're likely to adapt to it, but you don't realize up front that you will. We are bad at "affective forecasting,"[5] that is, predicting how we'll feel in the future. We grossly overestimate the intensity and the duration of our emotional reactions. Within a year, lottery winners and paraplegics have both (on average) returned most of the way to their baseline levels of happiness.[6] The lottery winner buys a new house and a new car, quits her boring job, and eats better food. She gets a kick out of the contrast with her former life, but within a few months the contrast blurs and the pleasure fades. The human mind is extraordinarily sensitive to *changes* in conditions, but not so sensitive to absolute levels. The winner's pleasure comes from rising in wealth, not from standing still at a high level, and after a few months the new comforts have become the new baseline of daily life. The winner takes them for granted and has no way to rise any further. Even worse: The money might damage her relationships. Friends, relatives, swindlers, and sobbing strangers swarm around lottery winners, suing them, sucking up to them, demanding a share of the wealth. (Remember the ubiquity of self-serving biases; everyone can find a reason to be owed something.) Lottery winners are so often harassed that many have to move, hide, end relationships, and finally turn to each other, forming lottery winner support groups to deal with their new difficulties.[7] (It should be noted, however, that nearly all lottery winners are still glad that they won.)

At the other extreme, the quadriplegic takes a huge happiness loss up front. He thinks his life is over, and it hurts to give up everything he once hoped for. But like the lottery winner, his mind is sensitive more to changes than to absolute levels, so after a few months he has begun adapting to his new situation and is setting more modest goals. He discovers that physical therapy can expand his abilities. He has nowhere to go but up, and each step gives him the pleasure of the progress principle. The physicist Stephen Hawking has been trapped in a shell of a body since his early twenties, when he was diagnosed with motor neurone disease. Yet he went on to solve major problems in cosmology, win many prizes, and write the best-selling science book of all time. During a recent interview in the *New York Times*, he was asked how he keeps his spirits up. He replied: "My expectations were reduced to zero when I was twenty-one. Everything since then has been a bonus."[8]

This is the adaptation principle at work: People's judgments about their present state are based on whether it is better or worse than the state to which they have become accustomed.[9] Adaptation is, in part, just a property of neurons: Nerve cells respond vigorously to new stimuli, but gradually they "habituate," firing less to stimuli that they have become used to. It is *change* that contains vital information, not steady states. Human beings, however, take adaptation to cognitive extremes. We don't just habituate, we recalibrate. We create for ourselves a world of targets, and each time we hit one we replace it with another. After a string of successes we aim higher; after a massive setback, such as a broken neck, we aim lower. Instead of following Buddhist and Stoic advice to surrender attachments and let events happen, we surround ourselves with goals, hopes, and expectations, and then feel pleasure and pain in relation to our progress.[10]

When we combine the adaptation principle with the discovery that people's average level of happiness is highly heritable,[11] we come to a startling possibility: In the long run, it doesn't much matter what happens to you. Good fortune or bad, you will always return to your happiness setpoint—your brain's default level of happiness—which was determined largely by your genes. In 1759, long before anyone knew about genes, Adam Smith reached the same conclusion:

> In every permanent situation, where there is no expectation of change, the mind of every man, in a longer or shorter time, returns to its natural and usual state of tranquility. In prosperity, after a certain time, it falls back to that state; in adversity, after a certain time, it rises up to it.[12]

If this idea is correct, then we are all stuck on what has been called the "hedonic treadmill."[13] On an exercise treadmill you can increase the speed all you want, but you stay in the same place. In life, you can work as hard as you want, and accumulate all the riches, fruit trees, and concubines you want, but you can't get ahead. Because you can't change your "natural and usual state of tranquility," the riches you accumulate will just raise your expectations and leave you no better off than you were before. Yet, not realizing the futility of our efforts, we continue to strive, all the while doing things that help us win at the game of life. Always wanting more than we have, we run and run and run, like hamsters on a wheel.

## AN EARLY HAPPINESS HYPOTHESIS

Buddha, Epictetus, and many other sages saw the futility of the rat race and urged people to quit. They proposed a particular happiness hypothesis: *Happiness comes from within, and it cannot be found by making the world conform to your desires.* Buddhism teaches that attachment leads inevitably to suffering and offers tools for breaking attachments. The Stoic philosophers of Ancient Greece, such as Epictetus, taught their followers to focus only on what they could fully control, which meant primarily their own thoughts and reactions. All other events—the gifts and curses of fortune—were externals, and the true Stoic was unaffected by externals.

Neither Buddha nor the Stoics urged people to withdraw into a cave. In fact, both doctrines have such enduring appeal precisely because they offer guidance on how to find peace and happiness while participating in a treacherous and ever-changing social world. Both doctrines are based on an empirical claim, a happiness hypothesis that asserts that striving to obtain goods and goals in the external world cannot bring you more than momentary happiness. You must work on your internal world. If the hypothesis is true, it has profound implications for how we should live our lives, raise our children, and spend our money. But is it true? It all depends on what kind of externals we are talking about.

The second biggest finding in happiness research, after the strong influence of genes upon a person's average level of happiness, is that most environmental and demographic factors influence happiness very little. Try to imagine yourself changing places with either Bob or Mary. Bob is thirty-five years old, single, white, attractive, and athletic. He earns $100,000 a year and lives in sunny Southern California. He is highly intellectual, and he spends his free time reading and going to museums. Mary and her husband live in snowy Buffalo, New York, where they earn a combined income of $40,000. Mary is sixty-five years old, black, overweight, and plain in appearance. She is highly sociable, and she spends her free time mostly in activities related to her church. She is on dialysis for kidney problems. Bob seems to have it all, and few readers of this book would prefer Mary's life to his. Yet if you had to bet on it, you should bet that Mary is happier than Bob.

What Mary has that Bob lacks are strong connections. A good marriage is one of the life-factors most strongly and consistently associated with happiness.[14] Part of this apparent benefit comes from "reverse correlation": Happiness causes marriage. Happy people marry sooner and stay married longer than people with a lower happiness setpoint, both because they are more appealing as dating partners and because they are easier to live with as spouses.[15] But much of the apparent benefit is a real and lasting benefit of dependable companionship, which is a basic need; we never fully adapt either to it or to its absence.[16] Mary also has religion, and religious people are happier, on average, than nonreligious people.[17] This effect arises from the social ties that come with participation in a religious community, as well as from feeling connected to something beyond the self.

What Bob has going for him is a string of objective advantages in power, status, freedom, health, and sunshine—all of which are subject to the adaptation principle. White Americans are freed from many of the hassles and indignities that affect black Americans, yet, on average, they are only very slightly happier.[18] Men have more freedom and power than women, yet they are not on average any happier. (Women experience more depression, but also more intense joy).[19] The young have so much more to look forward to than the elderly, yet ratings of life satisfaction actually rise slightly with age, up to age sixty-five, and, in some studies, well beyond.[20] People are often surprised to hear that the old are happier than the young because the old have so many more health problems, yet people adapt to most chronic health problems such as Mary's[21] (although ailments that grow progressively worse do reduce well-being, and a recent study finds that adaptation to disability is not, on average, complete).[22] People who live in cold climates expect people who live in California to be happier, but they are wrong.[23] People believe that attractive people are happier than unattractive people,[24] but they, too, are wrong.[25]

The one thing Bob does have going for him is wealth, but here the story is complicated. The most widely reported conclusion, from surveys done by psychologist Ed Diener,[26] is that within any given country, at the lowest end of the income scale money does buy happiness: People who worry every day about paying for food and shelter report significantly less well-being than those who don't. But once you are freed from basic needs and have entered the middle class, the relationship between wealth and happiness becomes smaller. The rich are happier on average than the middle class, but only by a little, and part of this relationship is reverse correlation: Happy people grow rich faster because, as in the marriage market, they are more appealing to others (such as bosses), and also because their frequent positive emotions help them to commit to projects, to work hard, and to invest in their futures.[27] Wealth itself has only a small direct effect on happiness because it so effectively speeds up the hedonic treadmill. For example, as the level of wealth has doubled or tripled in the last fifty years in many industrialized nations, the levels of happiness and satisfaction with life that people report have not changed, and depression has actually become more common.[28] Vast increases in gross domestic product led to improvements in the comforts of life—a larger home, more cars, televisions, and restaurant meals, better health and longer life—but these improvements became the normal conditions of life; all were adapted to and taken for granted, so they did not make people feel any happier or more satisfied.

These findings would have pleased Buddha and Epictetus—if, that is, they found pleasure in such external events as being proved right. As in their day, people today devote themselves to the pursuit of goals that won't make them happier, in the process neglecting the sort of inner growth and spiritual development that could bring lasting satisfaction. One of the most consistent lessons the ancient sages teach is to let go, stop striving, and choose a new path. Turn inwards, or toward God, but for God's sake stop trying to make the world conform to your will. The *Bhagavad Gita* is a Hindu treatise on nonattachment. In a section on "human devils," the god Krishna describes humanity's lower nature and the people who give in to it: "Bound by hundreds of fetters forged by hope, obsessed by anger and desire, they seek to build up wealth unjustly to satisfy their lusts."[29] Krishna then parodies the thinking of such a devil:

This have I gained today, this whim I'll satisfy; this wealth is mine and much more too will be mine as time goes on. He was an enemy of mine, I've killed him, and many another too I'll kill. I'm master here. I take my pleasure as I will. I'm strong and happy and successful.

Substitute "defeat" for "kill" and you have a pretty good description of the modern Western ideal, at least in some corners of the business world. So even if Bob were just as happy as Mary, if he has an arrogant, entitled attitude and treats people badly, his life would still be spiritually and aesthetically worse.

## THE HAPPINESS FORMULA

In the 1990s, the two big findings of happiness research (strong relation to genes, weak relation to environment) hit the psychological community hard, because they applied not just to happiness but to most aspects of personality. Psychologists since Freud had shared a nearly religious devotion to the idea that personality is shaped primarily by childhood environment. This axiom was taken on faith: The evidence for it consisted almost entirely of correlations—usually small ones—between what parents did and how their children turned out, and anyone who suggested that these correlations were caused by genes was dismissed as a reductionist. But as twin studies revealed the awesome reach of genes and the relative unimportance of the family environment that siblings share,[30] the ancient happiness hypothesis grew ever more plausible. Maybe there really is a set point[31] fixed into every brain, like a thermostat set forever to 58 degrees Fahrenheit (for depressives) or 75 degrees (for happy people)? Maybe the only way to find happiness therefore is to change one's own internal setting (for example, through meditation, Prozac, or cognitive therapy) instead of changing one's environment?

As psychologists wrestled with these ideas, however, and as biologists worked out the first sketch of the human genome, a more sophisticated understanding of nature and nurture began to emerge. Yes, genes explain far more about us than anyone had realized, but the genes themselves often turn out to be sensitive to environmental conditions.[32] And yes, each person has a characteristic level of happiness, but it now looks as though it's not so much a set *point* as a potential *range* or probability distribution. Whether you operate on the high or the low side of your potential range is determined by many factors that Buddha and Epictetus would have considered externals.

When Martin Seligman founded positive psychology in the late 1990s, one of his first moves was to bring together small groups of experts to tackle specific problems. One group was created to study the externals that matter for happiness. Three psychologists, Sonja Lyubomirsky, Ken Sheldon, and David Schkade, reviewed the available evidence and realized that there are two fundamentally different kinds of externals: the *conditions* of your life and the *voluntary activities* that you undertake.[33] Conditions include facts about your life that you can't change (race, sex, age, disability) as well as things that you can (wealth, marital status, where you live). Conditions are constant over time, at least during a period in your life, and so they are the sorts of things that you are likely to adapt to. Voluntary activities, on the other hand, are the things that you *choose* to do, such as meditation, exercise, learning a new skill, or taking a vacation. Because such activities must be chosen, and because most of them take effort and attention, they can't just disappear from your awareness the way conditions can. Voluntary activities, therefore, offer much greater promise for increasing happiness while avoiding adaptation effects.

One of the most important ideas in positive psychology is what Lyubomirsky, Sheldon, Schkade, and Seligman call the "happiness formula:"

$$H = S + C + V$$

The level of happiness that you actually experience (H) is determined by your biological set point (S) plus the conditions of your life (C) plus the voluntary activities (V) you do.[34] The challenge for positive psychology is to use the scientific method to find out exactly what kinds of C and V can push H up to the top of your potential range. The extreme biological version of the happiness hypothesis says that H = S, and that C and V don't matter. But we have to give Buddha and Epictetus credit for V because Buddha prescribed the "eightfold noble path" (including meditation and mindfulness), and Epictetus urged methods of thought to cultivate indifference (*apatheia*) to externals. So to test the wisdom of the sages properly we must examine this hypothesis: H = S + V, where V = voluntary or intentional activities that cultivate acceptance and weaken emotional attachments. If there are many conditions (C) that matter, and if there are a variety of voluntary activities beyond those aimed at nonattachment, then the happiness hypothesis of Buddha and Epictetus is wrong and people would be poorly advised simply to look within.

It turns out that there really are some external conditions (C) that matter. There are some changes you can make in your life that are not fully subject to the adaptation principle, and that might make you lastingly happier. It may be worth striving to achieve them.

*Noise.*   When I lived in Philadelphia, I learned a valuable lesson about real estate: If you must buy a house on a busy street, don't buy one within thirty yards of a traffic light. Every ninety-five seconds I had to listen to forty-two seconds of several people's musical selections followed by twelve seconds of engines revving, with an impatient honk thrown in once every fifteen cycles. I never got used to it, and when my wife and I were looking for a house in Charlottesville, I told our agent that if a Victorian mansion were being given away on a busy street, I would not take it. Research shows that people who must adapt to new and chronic sources of noise (such as when a new highway is built) never fully adapt, and even studies that find some adaptation still find evidence of impairment on cognitive tasks. Noise, especially noise that is variable or intermittent, interferes with concentration and increases stress.[35] It's worth striving to remove sources of noise in your life.

*Commuting.*   Many people choose to move farther away from their jobs in search of a larger house. But although people quickly adapt to having more space,[36] they don't fully adapt to the longer commute, particularly if it involves driving in heavy traffic.[37] Even after years of commuting, those whose commutes are traffic-filled still arrive at work with higher levels of stress hormones. (Driving under ideal conditions is, however, often enjoyable and relaxing.)[38] It's worth striving to improve your commute.

*Lack of control.*   One of the active ingredients of noise and traffic, the aspect that helps them get under your skin, is that you can't control them. In one classic study, David Glass and Jerome Singer exposed people to loud bursts of random noise. Subjects in one group were told they could terminate the noise by pressing a button, but they were asked not to press the button unless it was absolutely necessary. None of these subjects pressed the button, yet the belief that they had some form of control made the noise less distressing to them. In the second part of the experiment, the subjects who thought they had control were more persistent when working on difficult puzzles, but the subjects who had experienced noise without control gave up more easily.[39]

In another famous study, Ellen Langer and Judith Rodin gave benefits to residents on two floors of a nursing home—for example, plants in their rooms, and a movie screening one night a week. But on one floor, these benefits came with a sense of control: The residents were allowed to choose which plants they wanted, and they were responsible for watering them. They were allowed to choose as a group which night would be movie night. On the other floor, the same benefits were simply doled out: The nurses chose the plants and watered them; the nurses decided which night was movie night. This small manipulation had big effects: On the floor with increased control, residents were happier, more active, and more alert (as rated by the nurses, not just by the residents), and these benefits were still visible eighteen months later. Most amazingly, at the eighteen-month follow-up, residents of the floor given control had better health and half as many deaths (15 percent versus 30 percent).[40] In a review paper that Rodin and I wrote, we concluded that changing an institution's environment to increase the sense of control among its workers, students, patients, or other users was one of the most effective possible ways to increase their sense of engagement, energy, and happiness.[41]

*Shame.*    Overall, attractive people are not happier than unattractive ones. Yet, surprisingly, some improvements in a person's appearance do lead to lasting increases in happiness.[42] People who undergo plastic surgery report (on average) high levels of satisfaction with the process, and they even report increases in the quality of their lives and decreases in psychiatric symptoms (such as depression and anxiety) in the years after the operation. The biggest gains were reported for breast surgery, both enlargement and reduction. I think the way to understand the long-lasting effects of such seemingly shallow changes is to think about the power of shame in everyday life.

Young women whose breasts are much larger or smaller than their ideal often report feeling self-consciousness every day about their bodies. Many adjust their posture or their wardrobe in an attempt to hide what they see as a personal deficiency. Being freed from such a daily burden may lead to a lasting increase in self-confidence and well-being.

*Relationships.*   The condition that is usually said[43] to trump all others in importance is the strength and number of a person's relationships. Good relationships make people happy, and happy people enjoy more and better relationships than unhappy people.[44] This effect is so important and interesting that it gets its own chapter—the next one. For now, I'll just mention that conflicts in relationships—having an annoying office mate or roommate, or having chronic conflict with your spouse—is one of the surest ways to reduce your happiness. You never adapt to interpersonal conflict;[45] it damages every day, even days when you don't see the other person but ruminate about the conflict nonetheless.

There are many other ways in which you can increase your happiness by getting the conditions of your life right, particularly in relationships, work, and the degree of control you have over stressors. So in the happiness formula, C is real and some externals matter. Some things are worth striving for, and positive psychology can help identify them. Of course, Buddha would adapt fully to noise, traffic, lack of control and bodily deficiencies, but it has always been difficult, even in ancient India, for real people to become like Buddha. In the modern Western world, it is even harder to follow Buddha's path of nondoing and nonstriving. Some of our poets and writers in fact urge us to forswear that path and embrace action wholeheartedly: "It is vain to say that human beings ought to be satisfied with tranquility: they must have action; and they will make it if they cannot find it." (CHARLOTTE BRONTË, 1847)[46]

## FINDING FLOW

Not all action, however, will work. Chasing after wealth and prestige, for example, will usually backfire. People who report the greatest interest in attaining money, fame, or beauty are consistently found to be less happy, and even less healthy, than those who pursue less materialistic goals.[47] So what is the right kind of activity? What is V in the happiness formula?

The tool that helped psychologists answer that question is the "experience sampling method," invented by Mihalyi Csikszentmihalyi (pronounced "cheeks sent me high"), the Hungarian-born cofounder of positive psychology. In Csikszentmihalyi's studies,[48] people carry with them a pager that beeps several times a day. At each beep, the subject pulls out a small notebook and records what she is doing at that moment, and how much she is enjoying it. Through this "beeping" of thousands of people tens of thousands of times, Csikszentmihalyi found out what people really enjoy doing, not just what they *remember* having enjoyed. He discovered that there are two different kinds of enjoyment. One is physical or bodily pleasure. At meal times, people report the highest levels of happiness, on average. People really enjoy eating, especially in the company of others, and they hate to be interrupted by telephone calls (and perhaps Csikszentmihalyi's beeps) during meals, or (worst of all) during sex. But you can't enjoy physical pleasure all day long. By their very nature, food and sex satiate. To continue eating or having sex beyond a certain level of satisfaction can lead to disgust.[49]

Csikszentmihalyi's big discovery is that there is a state many people value even more than chocolate after sex. It is the state of total immersion in a task that is challenging yet closely matched to one's abilities. It is what people sometimes call "being in the zone." Csikszentmihalyi called it "flow" because it often feels like effortless movement: Flow happens, and you go with it. Flow often occurs during physical movement—skiing, driving fast on a curvy country road, or playing team sports. Flow is aided by music or by the action of other people, both of which provide a temporal structure for one's own behavior (for example, singing in a choir, dancing, or just having an intense conversation with a friend). And flow can happen during solitary creative activities, such as painting, writing, or photography. The keys to flow: There's a clear challenge that fully engages your attention; you have the skills to meet the challenge; and you get immediate feedback about how you are doing at each step (the progress principle). You get flash after flash of positive feeling with each turn negotiated, each high note correctly sung, or each brushstroke that falls into the right place. In the flow experience, elephant and rider are in perfect harmony. The elephant (automatic processes) is doing most of the work, running smoothly through the forest, while the rider (conscious thought) is completely absorbed in looking out for problems and opportunities, helping wherever he can.

Drawing on Csikszentmihalyi's work, Seligman proposes a fundamental distinction between pleasures and gratifications. Pleasures are "delights that have clear sensory and strong emotional components,"[50] such as may be derived from food, sex, backrubs, and cool breezes. Gratifications are activities that engage you fully, draw on your strengths, and allow you to lose self-consciousness. Gratifications can lead to flow. Seligman proposes that V (voluntary activities) is largely a matter of arranging your day and your environment to increase both pleasures and gratifications. Pleasures must be spaced to maintain their potency. Eating a quart of ice cream in an afternoon or listening to a new CD ten times in a row are good ways to overdose and deaden yourself to future pleasure. Here's where the rider has an important role to play: Because the elephant has a tendency to over-indulge, the rider needs to encourage it to get up and move on to another activity.

Pleasures should be both savored and varied. The French know how to do this: They eat many fatty foods, yet they end up thinner and healthier than Americans, and they derive a great deal more pleasure from their food by eating slowly and paying more attention to the food as they eat it.[51] Because they savor, they ultimately eat less. Americans, in contrast, shovel enormous servings of high-fat and high-carbohydrate food into their mouths while doing other things. The French also vary their pleasure by serving many small courses; Americans are seduced by restaurants that serve large portions. Variety is the spice of life because it is the natural enemy of adaptation. Super-sizing portions, on the other hand, maximizes adaptation. Epicurus, one of the few ancient philosophers to embrace sensual pleasure, endorsed the French way when he said that the wise man "chooses not the greatest quantity of food but the most tasty."[52]

One reason for the widespread philosophical wariness of sensual pleasure is that it gives no lasting benefit. Pleasure feels good in the moment, but sensual memories fade quickly, and the person is no wiser or stronger afterwards. Even worse, pleasure beckons people back for more, away from activities that might be better for them in the long run. But gratifications are different. Gratifications ask more of us; they challenge us and make us extend ourselves. Gratifications often come from accomplishing something, learning something, or improving something. When we enter a state of flow, hard work becomes effortless. We *want* to keep exerting ourselves, honing our skills, using our strengths. Seligman suggests that the key to finding your own gratifications is to know your own strengths.[53] One of the big accomplishments of positive psychology has been the development of a catalog of strengths. You can find out your strengths by taking an online test at www.authentichappiness.org.

Recently I asked the 350 students in my introductory psychology class to take the strengths test and then, a week later, to engage in four activities over a few days. One of the activities was to indulge the senses, as by taking a break for ice cream in the middle of the afternoon, and then savoring the ice cream. This activity was the most enjoyable at the time; but, like all pleasures, it faded quickly. The other three activities were potential gratifications: Attend a lecture or class that you don't normally go to; perform an act of kindness for a friend who could use some cheering up; and write down the reasons you are grateful to someone and later call or visit that person to express your gratitude. The least enjoyable of the four activities was going to a lecture—except for those whose strengths included curiosity and love of learning. They got a lot more out of it. The big finding was that people experienced longer-lasting improvements in mood from the kindness and gratitude activities than from those in which they indulged themselves. Even though people were most nervous about doing the kindness and gratitude activities, which required them to violate social norms and risk embarrassment, once they actually did the activities they felt better for the rest of the day. Many students even said their good feelings continued on into the next day—which nobody said about eating ice cream. Furthermore, these benefits were most pronounced for those whose strengths included kindness and gratitude.

So V (voluntary activity) is real, and it's not just about detachment. You can increase your happiness if you use your strengths, particularly in the

service of strengthening connections—helping friends, expressing gratitude to benefactors. Performing a random act of kindness every day could
get tedious, but if you know your strengths and draw up a list of five activities that engage them, you can surely add at least one gratification to every
day. Studies that have assigned people to perform a random act of kindness
every week, or to count their blessings regularly for several weeks, find
small but sustained increases in happiness.[54] So take the initiative! Choose
your own gratifying activities, do them regularly (but not to the point of te-
dium), and raise your overall level of happiness.


## MISGUIDED PURSUITS

An axiom of economics is that people pursue their interests more or less rationally, and that's what makes markets work—Adam Smith's "invisible hand"
of self-interest. But in the 1980s, a few economists began studying psychology and messing up the prevailing models. Leading the way was the Cornell
economist Robert Frank, whose 1987 book *Passions Within Reason* analyzed
some of the things people do that just don't fit into economic models of pure
self-interest—such as tipping in restaurants when far from home, seeking
costly revenge, and staying loyal to friends and spouses when better opportunities come along. Frank argued that these behaviors make sense only as
products of moral emotions (such as love, shame, vengeance, or guilt), and
these moral emotions make sense only as products of evolution. Evolution
seems to have made us "strategically irrational" at times for our own good;
for example, a person who gets angry when cheated, and who will pursue
vengeance regardless of the cost, earns a reputation that discourages would-
be cheaters. A person who pursued vengeance only when the benefits out-
weighed the costs could be cheated with impunity in many situations.

In his more recent book, *Luxury Fever*,[55] Frank used the same approach
to understand another kind of irrationality: the vigor with which people
pursue many goals that work against their own happiness. Frank begins
with the question of why, as nations rise in wealth, their citizens become
no happier, and he considers the possibility that once basic needs are met,

money simply cannot buy additional happiness. After a careful review of the evidence, however, Frank concludes that those who think money can't buy happiness just don't know where to shop. Some purchases are much less subject to the adaptation principle. Frank wants to know why people are so devoted to spending money on luxuries and other goods, to which they adapt completely, rather than on things that would make them lastingly happier. For example, people would be happier and healthier if they took more time off and "spent" it with their family and friends, yet America has long been heading in the opposite direction. People would be happier if they reduced their commuting time, even if it meant living in smaller houses, yet American trends are toward ever larger houses and ever longer commutes. People would be happier and healthier if they took longer vacations, even if that meant earning less, yet vacation times are shrinking in the United States, and in Europe as well. People would be happier, and in the long run wealthier, if they bought basic, functional appliances, automobiles, and wristwatches, and invested the money they saved for future consumption; yet, Americans in particular spend almost everything they have—and sometimes more—on goods for present consumption, often paying a large premium for designer names and superfluous features.

Frank's explanation is simple: Conspicuous and inconspicuous consumption follow different psychological rules. Conspicuous consumption refers to things that are visible to others and that are taken as markers of a person's relative success. These goods are subject to a kind of arms race, where their value comes not so much from their objective properties as from the statement they make about their owner. When everyone wore Timex watches, the first person in the office buy a Rolex stood out. When everyone moved up to Rolex, it took a $20,000 Patek Philip to achieve high status, and a Rolex no longer gave as much satisfaction. Conspicuous consumption is a zero-sum game: Each person's move up devalues the possessions of others. Furthermore, it's difficult to persuade an entire group or subculture to ratchet down, even though everyone would be better off, on average, if they all went back to simple watches. Inconspicuous consumption, on the other hand, refers to goods and activities that are valued for themselves, that are usually consumed more privately, and that are not bought for the purpose of achieving status. Because Americans, at least, gain no prestige from taking the longest vacations or having the shortest commutes, these inconspicuous consumables are not subject to an arms race.

Just try this thought experiment. Which job would you rather have: one in which you earned $90,000 a year and your coworkers earned on average $70,000, or one in which you earned $100,000 but your coworkers earned on average $150,000? Many people choose the first job, thereby revealing that relative position is worth at least $10,000 to them. Now try this one: Would you rather work for a company that gave you two weeks of vacation a year, but other employees were given, on average, only one; or would you prefer a company that gave you four weeks of vacation a year, but other employees were given, on average, six? The great majority of people choose the longer absolute time.[56] Time off is inconspicuous consumption, although people can easily turn a vacation into conspicuous consumption by spending vast amounts of money to impress others instead of using the time to rejuvenate themselves.

Frank's conclusions are bolstered by recent research on the benefits of "doing versus having." The psychologists Leaf van Boven and Tom Gilovich asked people to think back to a time when they spent more than a hundred dollars with the intention of increasing their happiness and enjoyment. One group of subjects was asked to pick a material possession; the other was asked to pick an experience or activity they had paid for. After describing their purchases, subjects were asked to fill out a questionnaire. Those who described buying an experience (such as a ski trip, a concert, or a great meal) were happier when thinking about their purchase, and thought that their money was better spent, than those who described buying a material object (such as clothing, jewelry, or electronics).[57] After conducting several variations of this experiment with similar findings each time, Van Boven and Gilovich concluded that experiences give more happiness in part because they have greater social value: Most activities that cost more than a hundred dollars are things we do *with* other people, but expensive material possessions are often purchased in part to *impress* other people. Activities connect us to others; objects often separate us.

So now you know where to shop. Stop trying to keep up with the Joneses. Stop wasting your money on conspicuous consumption. As a first step, work less, earn less, accumulate less, and "consume" more family time, vacations, and other enjoyable activities. The Chinese sage Lao Tzu warned people to make their own choices and not pursue the material objects everyone else was pursuing:

> Racing and hunting madden the mind.
> Precious things lead one astray.
> Therefore the sage is guided by what he feels and not by what he sees.
> He lets go of that and chooses this.[58]

Unfortunately, letting go of one thing and choosing another is difficult if the elephant wraps his trunk around the "precious thing" and refuses to let go. The elephant was shaped by natural selection to win at the game of life, and part of its strategy is to impress others, gain their admiration, and rise in relative rank. *The elephant cares about prestige, not happiness,*[59] and it looks eternally to others to figure out what is prestigious. The elephant will pursue its evolutionary goals even when greater happiness can be found elsewhere. If everyone is chasing the same limited amount of prestige, then all are stuck in a zero-sum game, an eternal arms race, a world in which rising wealth does not bring rising happiness. The pursuit of luxury goods is a happiness trap; it is a dead end that people race toward in the mistaken belief that it will make them happy.

Modern life has many other traps. Here's some bait. Of the following words, pick the one that is most appealing to you: *constraint, limit, barrier, choice*. Odds are you chose *choice,* because the first three gave you a flash of negative affect (remember the like-o-meter). Choice and its frequent associate freedom are unquestioned goods of modern life. Most people would rather shop at a supermarket that stocks ten items in each food category than at a small store that stocks just two. Most people would prefer to invest their retirement savings through a company that offers forty funds than one that offers four. Yet, when people are actually given a larger array of choices—for example, an assortment of thirty (rather than six) gourmet chocolates from which to choose—they are less likely to make a choice; and if they do, they are less satisfied with it.[60] The more choices there are, the more you expect to find a perfect fit; yet, at the same time, the larger the array, the less likely it becomes that you picked the best item. You leave the store less confident in your choice, more likely to feel regret, and more likely to think about the options you didn't choose. If you can avoid making a choice, you are more likely to do so. The psychologist Barry Schwartz calls this the "paradox of choice":[61] We value choice and put ourselves in situations of choice, even though choice often undercuts our happiness. But Schwartz and his colleagues[62] find that the paradox mostly applies to people they call "maximizers"—those who habitually try to evaluate all the options, seek out more information, and make the best choice (or "maximize their utility," as economists would say). Other people—"satisficers"—are more laid back about choice. They evaluate an array of options until they find one that is good enough, and then they stop looking. Satisficers are not hurt by a surfeit of options. Maximizers end up making slightly better decisions than satisficers, on average (all that worry and information-gathering does help), but they are less happy with their decisions, and they are more inclined to depression and anxiety.

In one clever study,[63] maximizers and satisficers were asked to solve anagrams while sitting next to another subject (really a co-experimenter) who was solving them either much faster or much slower. Satisficers were relatively unfazed by the experience. Their ratings of their own ability, and of how much they enjoyed the study, were barely affected by what the other subject did. But maximizers were thrown for a loop when the other subject was faster than they were. They later reported lower estimates of their own abilities and higher levels of negative emotions. (Being paired with a slower peer didn't have much effect—another instance of negative events being stronger than positive). The point here is that maximizers engage in more social comparison, and are therefore more easily drawn into conspicuous consumption. Paradoxically, maximizers get less pleasure per dollar they spend.

Modern life is full of traps. Some of these traps are set by marketers and advertisers who know just what the elephant wants—and it isn't happiness.

## THE HAPPINESS HYPOTHESIS RECONSIDERED

When I began writing this book, I thought that Buddha would be a strong contender for the "Best Psychologist of the Last Three Thousand Years" award. To me, his diagnosis of the futility of striving felt so right, his promise of tranquility so alluring. But in doing research for the book, I began to think that Buddhism might be based on an overreaction, perhaps even an error. According to legend,[64] Buddha was the son of a king in northern India. When he was born (as Siddhartha Gautama), the king heard a prophecy that his son was destined to leave, to go into the forest and turn his back on the kingdom. So as the boy grew into adulthood, his father tried to tie him down with sensual pleasures and hide from him anything that might disturb his mind. The young prince was married to a beautiful princess and raised on the upper floors of the palace, surrounded by a harem of other beautiful women. But he grew bored (the adaptation principle) and curious about the world outside. Eventually, he prevailed upon his father to let him go for a chariot ride. On the morning of the ride, the king ordered that all people who were old, sick, or crippled were to retreat indoors. Yet one old man remained on the road, and the prince saw him. The prince asked his chariot driver to explain the odd-looking creature, and the driver told him that everyone grows old. Stunned, the young prince returned to his palace. On the next day's excursion, he saw a sick man, his body hobbled by disease. More explanation, more retreating to the palace. On the third day, the prince saw a corpse being carried through the streets. This was the last straw. Upon discovering that old age, disease, and death are the destiny of all people, the prince cried, "Turn back the chariot! This is no time or place for pleasure excursions. How could an intelligent person pay no heed at a time of disaster, when he knows of his impending destruction?"[65] The prince then left his wife, his harem, and, as prophesied, his royal future. He went into the forest and began his journey to enlightenment. After his enlightenment, Buddha[66] (the "awakened one") preached that life is suffering, and that the only way to escape this suffering is by breaking the attachments that bind us to pleasure, achievement, reputation, and life.

But what would have happened if the young prince had actually descended from his gilded chariot and talked to the people he assumed were so miserable? What if he had interviewed the poor, the elderly, the crippled, and the sick? One of the most adventurous young psychologists, Robert Biswas-Diener (son of the happiness pioneer Ed Diener), has done just that. He has traveled the world interviewing people about their lives and how satisfied they are with them. Wherever he goes, from Greenland to Kenya to California, he finds that most people (with the exception of homeless people) are more satisfied than dissatisfied with their lives.[67] He even interviewed sex workers in the slums of Calcutta, forced by poverty to sell their bodies and sacrifice their futures to disease. Although these women were substantially less satisfied with their lives than was a comparison group of college students in Calcutta, they still (on average) rated their satisfaction with each of twelve specific aspects of their lives as more satisfied than dissatisfied, or else as neutral (neither satisfied nor dissatisfied). Yes, they suffered privations that seem to us in the West unbearable, but they also had close friends with whom they spent much of their time, and most of them stayed in touch with their families. Biswas-Diener concludes that "while the poor of Calcutta do not lead enviable lives, they do lead meaningful lives. They capitalize on the non-material resources available to them and find satisfaction in many areas of their lives."[68] Like quadriplegics, the elderly, or any other class of people the young Buddha might have pitied, the lives of these prostitutes are much better from the inside than they seem from the outside.

Another reason for Buddha's emphasis on detachment may have been the turbulent times he lived in: Kings and city-states were making war, and people's lives and fortunes could be burned up overnight. When life is unpredictable and dangerous (as it was for the Stoic philosophers, living under capricious Roman emperors), it might be foolish to seek happiness by controlling one's external world. But now it is not. People living in wealthy democracies can set long-term goals and expect to meet them. We are immunized against disease, sheltered from storms, and insured against fire, theft, and collision. For the first time in human history, most people (in wealthy countries) will live past the age of seventy and will not see any of their children die before them. Although all of us will get unwanted surprises along the way, we'll adapt and cope with nearly all of them, and many of us will believe we are better off for having suffered. So to cut off all attachments, to shun the pleasures of sensuality and triumph in an effort to escape the pains of loss and defeat—this now strikes me as an inappropriate response to the inevitable presence of *some* suffering in every life.

Many Western thinkers have looked at the same afflictions as Buddha—sickness, aging, and mortality—and come to a very different conclusion from his: Through passionate attachments to people, goals, and pleasures, life must be lived to the fullest. I once heard a talk by the philosopher Robert Solomon, who directly challenged the philosophy of nonattachment as an affront to human nature.[69] The life of cerebral reflection and emotional indifference (*apatheia*) advocated by many Greek and Roman philosophers and that of calm nonstriving advocated by Buddha are lives designed to avoid passion, and a life without passion is not a human life. Yes, attachments bring pain, but they also bring our greatest joys, and there is value in the very variation that the philosophers are trying to avoid. I was stunned to hear a philosopher reject so much of ancient philosophy, but I was also inspired in a way that I had never been as an undergraduate student of philosophy. I walked out of the lecture hall feeling that I wanted to do something then and there to embrace life.

Solomon's message was unorthodox in philosophy, but it is common in the work of romantic poets, novelists, and nature writers: "We do not live but a quarter part of our life—why do we not let on the flood—raise the gates—& set our wheels in motion—He that hath ears to hear let him hear. Employ your senses." (HENRY DAVID THOREAU, 1851)[70]

Even a future justice of the U.S. Supreme Court—a body devoted to reason—issued this opinion: "I think that, as life is action and passion, it is required of a man that he should share the passion and action of his time at peril of being judged not to have lived." (OLIVER WENDELL HOLMES, JR., 1884)[71]

Buddha, Lao Tzu, and other sages of the East discovered a path to peace and tranquility, the path of letting go. They told us how to follow the path using meditation and stillness. Millions of people in the West have followed, and although few, if any, have reached Nirvana, many have found some degree of peace, happiness, and spiritual growth. So I do not mean to question the value or relevance of Buddhism in the modern world, or the importance of working on yourself in an effort to find happiness. Rather, I would like to suggest that the happiness hypothesis be extended—for now—into a yin-yang formulation: *Happiness comes from within, and happiness comes from without.* (In chapter 10, I'll suggest a further refinement of the hypothesis.)

To live both the yin and the yang, we need guidance. Buddha is history's most perceptive guide to the first half; he is a constant but gentle reminder of the yin of internal work. But I believe that the Western ideal of action, striving, and passionate attachment is not as misguided as Buddhism suggests. We just need some balance (from the East) and some specific guidance (from modern psychology) about what to strive for.

# 6

## Love and Attachments

*No one can live happily who has regard to himself alone and transforms everything into a question of his own utility; you must live for your neighbour, if you would live for yourself.*

— SENECA1

*No man is an island, entire of itself; every man is a piece of the continent, a part of the main.*

— JOHN DONNE2

IN 1931, AT THE AGE of four, my father was diagnosed with polio. He was immediately put into an isolation room at the local hospital in Brooklyn, New York. There was no cure and no vaccine for polio at that time, and city dwellers lived in fear of its spread. For several weeks my father had no human contact, save for an occasional visit by a masked nurse. His mother came to see him every day, but that's all she could do—wave to him and try to talk to him through the glass pane on the door. My father remembers calling out to her, begging her to come in. It must have broken her heart, and one day she ignored the rules and went in. She was caught and sternly reprimanded. My father recovered with no paralysis, but this image has always stayed with me: a small boy alone in a room, gazing at his mother through a pane of glass.

My father had the bad luck to be born at the confluence point of three big ideas. The first was germ theory, proposed in the 1840s by Ignaz Semmelweis and incorporated into hospitals and homes with gradually increasing ferocity over the next century. When they began to collect statistics from orphanages and foundling homes in the 1920s, pediatricians came to fear germs above all else. As far back as records went, they showed that most children dropped off at foundling homes died within one year. In 1915, a New York physician, Henry Chapin, reported to the American Pediatric Society that out of the ten foundling homes he had examined, in all but one of them *all* the children had died before their second birthday.[3] As pediatricians came to grips with the deadly effects of institutions on young children, they reacted in a logical way by launching a crusade against germs. It became a priority in orphanages and hospitals to isolate children as much as possible in clean cubicles to prevent them from infecting each other. Beds were separated, dividers were placed between beds, nurses retreated behind masks and gloves, and mothers were scolded for violating quarantine.

The other two big ideas were psychoanalysis and behaviorism. These two theories agreed on very little, but they both agreed that the infant's attachment to its mother is based on milk. Freud thought that the infant's libido (desire for pleasure) is first satisfied by the breast, and therefore the infant develops its first attachment (psychological need) to the breast. Only gradually does the child generalize that desire to the woman who owns the breast. The behaviorists didn't care about libido, but they, too, saw the breast as the first reinforcer, the first reward (milk) for the first behavior (sucking). The heart of behaviorism, if it had one, was conditioning—the idea that learning occurs when rewards are *conditional* upon behaviors. Unconditional love— holding, nuzzling, and cuddling children for no reason—was seen as the surest way to make children lazy, spoiled, and weak. Freudians and behaviorists were united in their belief that highly affectionate mothering damages children, and that scientific principles could improve child rearing. Three years before my father entered the hospital, John Watson, the leading American behaviorist (in the years before B. F. Skinner), published the best-seller *Psychological Care of Infant and Child*.[4] Watson wrote of his dream that one day babies would be raised in baby farms, away from the corrupting influences of parents. But until that day arrived, parents were urged to use behaviorist techniques to rear strong children: Don't pick them up when they cry, don't cuddle or coddle them, just dole out benefits and punishments for each good and bad action.

How could science have gotten it so wrong? How could doctors and psychologists not have seen that children need love as well as milk? This chapter is about that need—the need for other people, for touch, and for close relationships. No man, woman, or child is an island. Scientists have come a long way since John Watson, and there is now a much more humane science of love. The story of this science begins with orphans and rhesus monkeys and ends with a challenge to the dismal view of love held by many of the ancients, East and West. The heroes of this story are two psychologists who rejected the central tenets of their training: Harry Harlow and John Bowlby. These two men knew that something was missing in behaviorism and in psychoanalysis, respectively. Against great odds they changed their fields, they humanized the treatment of children, and they made it possible for science to greatly improve upon the wisdom of the ancients.

## To Have and to Hold

Harry Harlow[5] earned his Ph.D. in 1930 at Stanford, where he wrote his dissertation on the feeding behavior of baby rats. He took a job at the University of Wisconsin, where he found himself overwhelmed with teaching and undersupplied with research subjects—he had no lab space, no rats, no way to perform the experiments he was expected to publish. Out of desperation, Harlow took his students to the little zoo in Madison, Wisconsin, which had a small number of primates. Harlow and his first graduate student, Abe Maslow, couldn't run controlled experiments using so few animals. They were forced instead to observe, to keep their minds open, and to learn from species closely related to human beings. And one of the first things they saw was curiosity. The apes and monkeys liked to solve puzzles (the humans gave them tests to measure physical dexterity and intelligence), and would work at tasks for what seemed to be the sheer pleasure of it. Behaviorism, in contrast, said that animals will only do what they have been reinforced for doing.

Harlow sensed he had found a flaw in behaviorism, but he couldn't prove it with anecdotes from the local zoo. He desperately wanted a lab in which to study primates, not rats, so he built one himself—literally built it, in the shell of an abandoned building, with the help of his students. In that makeshift lab, for the next thirty years, Harlow and his students infuriated behaviorists by demonstrating with ever more precision that monkeys are curious, intelligent creatures who like to figure things out. They follow the laws of reinforcement to some degree, as do humans, but there is much more going on in a monkey brain than the brain of a behaviorist could grasp. For example, giving monkeys raisins as a reward for each correct step in solving a puzzle (such as opening a mechanical latch with several moving parts) actually interferes with the solving, because it distracts the monkeys.[6] They enjoy the task for its own sake.

As Harlow's lab grew, he faced perennial shortages of monkeys. They were hard to import and, when they arrived they were often sick, bringing a stream of new infections into the lab. In 1955, Harlow conceived the bold idea of starting his own breeding colony of rhesus monkeys. Nobody had ever created a self-sustaining breeding colony of monkeys in the United States, let alone in the cold climate of Wisconsin, but Harlow was undeterred. He allowed his rhesus monkeys to mate, and then he took away the children within hours of their birth—to save them from infections in the crowded lab. After much experimentation, he and his students created an artificial baby formula full of nutrients and antibiotics. They found the optimum pattern of feeding, light and dark cycles, and temperature. Each baby was raised in its own cage, safe from disease. Harlow had in a way realized Watson's dream of a baby farm, and the crop grew large and healthy-looking. But when the farm-raised monkeys were brought into the company of others, they were stunned and unnerved. They never developed normal social or problem-solving skills, so they were useless for experiments. Harlow and his students were stumped. What had they forgotten?

The clue was in plain sight, clutched in the monkeys' hands, until finally a grad student, Bill Mason, noticed it: diapers. The cages in the baby hatchery were sometimes lined with old diapers to provide bedding material and protect the babies from the cold floor. The monkeys clung to the diapers, especially when they were afraid, and took them along when they were carried to new cages. Mason proposed a test to Harlow: Let's expose some young monkeys to a bundle of cloth and a bundle of wood. Let's see whether the monkeys just need to hold on to something, anything, or whether there's something special about the softness of the cloth. Harlow loved the idea, and, as he thought it over, he saw an even grander question: Were the diapers really substitutes for mothers? Did the monkeys have an innate need to hold and be held, a need that was utterly starved in the baby farm? If so, how could he prove it? Harlow's proof became one of the most famous experiments in all of psychology.

Harlow put the milk hypothesis to a direct test. He created two kinds of surrogate mother, each one a cylinder about the size of an adult female rhesus monkey, complete with a wooden head that had eyes and a mouth. One kind was made of wire mesh, the other was covered with a layer of foam and then a layer of soft terrycloth. Each of eight baby rhesus monkeys was raised alone in a cage with two surrogate mothers, one of each kind. For four of the monkeys, milk was delivered only from a tube coming through the chest of the wire mother. For the other four, the tube came through the chest of the cloth mother. If Freud and Watson were right that milk was the cause of attachment, the monkeys should attach to their milk givers. But that's not what happened. All the monkeys spent nearly all their time clinging to, climbing on, and pushing themselves into the soft folds of the cloth mother. Harlow's experiment[7] is so elegant and so convincing that you don't need to see statistics to understand the results. You just need to see the famous photo, now included in every introductory psychology book, in which a baby monkey clings to the cloth mother with its hind legs while stretching over to feed from the tube protruding from the wire mother.

Harlow argued that "contact comfort" is a basic need that young mammals have for physical contact with their mother. In the absence of a real mother, young mammals will seek out whatever feels most like a mother. Harlow chose the term carefully, because the mother, even a cloth mother, provides comfort when it is most needed, and that comfort comes mostly from direct contact.

Displays of familial love often move people to tears, and Deborah Blum's wonderful biography of Harlow, *Love at Goon Park*,[8] is full of touching expressions of familial love. It is an uplifting story, ultimately, but along the way it is full of sadness and unrequited love. The cover of the book, for example, shows a picture of a young rhesus monkey alone in a cage, gazing at its cloth "mother" through a pane of glass.

## LOVE CONQUERS FEAR

John Bowlby's life followed an entirely different path from Harlow's, even though it led, ultimately, to the same discovery.[9] Bowlby was an English aristocrat, raised by a nanny, and sent to boarding school. He studied medicine and became a psychoanalyst, but during his early training years, he did some volunteer work that shaped the rest of his career. He worked at two homes for maladjusted children, many of whom had no real contact with their parents. Some were aloof and uncommunicative; others were hopelessly clingy, following him around anxiously if he paid the slightest attention to them. After serving in World War II, Bowlby returned to England to run the children's clinic in a hospital. He began to do research on how separation from parents affects children. Europe at that time had just experienced more parent-child separations than had any place in all of human history. The war had created vast numbers of orphans, refugees, and children sent away to the countryside for their own safety. The new World Health Organization commissioned Bowlby to write a report on the best way to deal with these children. Bowlby toured hospitals and orphanages, and his report, published in 1951, was a passionate argument against prevailing notions that separation and isolation are harmless, and that biological needs such as nutrition are paramount. Children need love to develop properly, he argued; children need mothers.

Throughout the 1950s, Bowlby developed his ideas and weathered the scorn of psychoanalysts such as Anna Freud and Melanie Klein, whose theories (about libido and breasts) he contradicted. He had the good luck to meet a leading ethologist of the day, Robert Hinde, who taught him about new research on animal behavior. Konrad Lorenz, for example, had demonstrated that ducklings, ten to twelve hours after they hatch, will lock onto whatever duck-sized thing moves around in their environment and then follow it around for months.[10] In nature this thing is always mom, but in Lorenz's demonstrations, anything he moved around worked—even his own boots (with him in them). This visual "imprinting" mechanism is quite different from what happens in people, but once Bowlby began to think about how evolution creates mechanisms to make sure that mothers and children stay together, the way was open for an entirely new approach to human parent-child relationships. There's no need to derive the bond from milk, reinforcement, libido, or anything else. Rather, the attachment of mother and child is so enormously important for the survival of the child that a dedicated system is built into mother and child in all species that rely on maternal care. As Bowlby began to pay more attention to animal behavior, he saw many similarities between the behaviors of baby monkeys and baby humans: clinging, sucking, crying when left behind, following whenever possible. All these behaviors functioned in other primates to keep the child close to mom, and all were visible in human children, even the "pick me up" signal of upstretched arms.

In 1957, Hinde learned about Harlow's not-yet-published cloth-mother studies and told Bowlby, who wrote to Harlow and later visited him in Wisconsin. The two men became great allies and supporters of each other. Bowlby, the great theorist, created the framework that has unified most subsequent research on parent-child relations; and Harlow, the great experimentalist, provided the first irrefutable lab demonstrations of the theory.

Bowlby's grand synthesis is called attachment theory.[11] It borrows from the science of cybernetics—the study of how mechanical and biological systems can regulate themselves to achieve preset goals while the environment around and inside them changes. Bowlby's first metaphor was the simplest cybernetic system of all—a thermostat that turns on a heater when the temperature drops below a set point.

Attachment theory begins with the idea that two basic goals guide children's behavior: safety and exploration. A child who stays safe survives; a child who explores and plays develops the skills and intelligence needed for adult life. (This is why all mammal babies play; and the larger their frontal cortex, the more they need to play).[12] These two needs are often opposed, however, so they are regulated by a kind of thermostat that monitors the level of ambient safety. When the safety level is adequate, the child plays and explores. But as soon as it drops too low, it's as though a switch were thrown and suddenly safety needs become paramount. The child stops playing and moves toward mom. If mom is unreachable, the child cries, and with increasing desperation; when mom returns, the child seeks touch, or some other reassurance, before the system can reset and play can resume. This is an instance of the "design" principle I discussed in chapter 2: opposing systems push against each other to reach a balance point. (Fathers make perfectly good attachment figures, but Bowlby focused on mother-child attachments, which usually get off to a faster start.)

If you want to see the system in action, just try engaging a two-year-old in play. If you go to a friend's house and meet her child for the first time, it should take only a minute. The child feels secure in his familiar surroundings, and his mother functions as what Bowlby called a "secure base"—an attachment figure whose presence guarantees safety, turns off fear, and thereby enables the explorations that lead to healthy development. But if your friend brings her son over to *your* house for the first time, it will take longer. You'll probably have to walk around your friend just to find the little head hiding behind her thighs. And then, if you succeed in starting a game—making faces at him to make him laugh, perhaps—just watch what happens when his mother goes to the kitchen to get a glass of water. The thermostat clicks, the game ends, and your play partner scampers off to the kitchen, too. Harlow had shown all the same behavior in monkeys.[13] Young monkeys placed with their cloth mother in the center of an open room full of toys eventually climbed down from mom to explore, but they returned often to touch her and reconnect. If the cloth mother was removed from the room, all play stopped and frantic screaming ensued.

When children are separated from their attachment figures for a long time, as in a hospital stay, they quickly descend into passivity and despair. When they are denied a stable and enduring attachment relationship (raised, for example, by a succession of foster parents or nurses), they are likely to be damaged for life, Bowlby said. They might become the aloof loners or hopeless clingers that Bowlby had seen in his volunteer work. Bowlby's theory directly contradicted Watson as well as the Freuds (Sigmund and Anna): If you want your children to grow up to be healthy and independent, you should hold them, hug them, cuddle them, and love them. Give them a secure base and they will explore and then conquer the world on their own. The power of love over fear was well expressed in the New Testament: "There is no fear in love, but perfect love casts out fear" (I JOHN 4:18).

## THE PROOF IS IN THE PARTING

If you're going to contradict the prevailing wisdom of your day, you'd better have darn good evidence. Harlow's studies were darn good, but skeptics claimed they didn't apply to people. Bowlby needed more proof, and he got it from a Canadian woman who happened to answer an ad he placed for a research assistant in 1950. Mary Ainsworth, who had moved to London with her husband, spent three years working with Bowlby on his early studies of hospitalized children. When her husband took an academic job in Uganda, Ainsworth went with him again and took advantage of the opportunity to make careful observations of children in Ugandan villages. Even in a culture where women share mothering duties for all the children in the extended family household, Ainsworth observed a special bond between a child and his own mother. The mother was much more effective as a secure base than were other women. Ainsworth then moved to the Johns Hopkins University in Baltimore, and after that to the University of Virginia, where she thought about how to test Bowlby's ideas, and her own, about the mother-child relationship.

In Bowlby's cybernetic theory, the action is in the changes. You can't just watch a child play; you have to look at how the exploration and safety goals shift in response to changing conditions. So Ainsworth developed a little drama, later called the "Strange Situation," and cast the child in the starring role.[14] In essence, she re-created the experiments in which Harlow had placed monkeys in an open room with novel toys. In the first scene, the mother and her child enter a comfortable room, full of toys. Most children in the experiment soon crawl or toddle off to explore. In scene two, a friendly woman enters, talks with the mother for a few minutes, and then joins the child in play. In scene three, the mother gets up and leaves the child alone for a few minutes with the stranger. In scene four, she returns and the stranger leaves. In scene five, the mother leaves again, and the child is all alone in the room. In scene six, the stranger returns; and in scene seven, the mother returns for good. The play is designed to ratchet up the child's stress level in order to see how the child's attachment system manages the scene changes. Ainsworth found three common patterns of managing.

In about two-thirds of American children, the system does just what Bowlby said it should, that is, shift smoothly between play and security-seeking as the situation changes. Children following this pattern, called "secure" attachment, reduce or stop their play when their mothers leave, and then show anxiety, which the stranger cannot fully relieve. In the two scenes where mom returns, these children show delight, often moving toward her or touching her to reestablish contact with their secure base; but then they quickly settle down and return to play. In the other third of children, the scene changes are more awkward; these children have one of two types of insecure attachment. The majority of them don't seem to care very much whether mom comes or goes, although subsequent physiological research showed that they are indeed distressed by the separation. Rather, these children seem to be suppressing their distress by trying to manage it on their own instead of relying upon mom for comfort. Ainsworth called this pattern "avoidant" attachment. The remaining children, about 12 percent in the United States, are anxious and clingy throughout the study. They become extremely upset when separated from mom, they sometimes resist her efforts to comfort them when she returns, and they never fully settle down to play in the unfamiliar room. Ainsworth called this pattern "resistant."[15]

Ainsworth first thought these differences were caused entirely by good or bad mothering. She observed mothers at home and found that those who were warm and highly responsive to their children were most likely to have children who showed secure attachment in the strange situation. These children had learned that they could count on their mothers, and were therefore the most bold and confident. Mothers who were aloof and unresponsive were more likely to have avoidant children, who had learned not to expect much help and comfort from mom. Mothers whose responses were erratic and unpredictable were more likely to have resistant children, who had learned that their efforts to elicit comfort sometimes paid off, but sometimes not.

But whenever I hear about correlations between mother and child, I'm skeptical. Twin studies almost always show that personality traits are due more to genetics than to parenting.[16] Maybe it's just that happy women, those who won the cortical lottery, are warm and loving, and they pass on their happy genes to their children, who then show up as securely attached. Or maybe the correlation runs in reverse: Children do have stable inborn temperaments[17]—sunny, cranky, or anxious—and the sunny ones are just so much fun that their mothers *want* to be more responsive. My skepticism is bolstered by the fact that studies done after Ainsworth's home study have generally found only small correlations between mothers' responsiveness and the attachment style of their children.[18] On the other hand, twin studies have found that genes play only a small role in determining attachment style.[19] So now we have a real puzzle, a trait that correlates weakly with mothering and weakly with genes. Where does it come from?

Bowlby's cybernetic theory forces us to think outside the usual nature-nurture dichotomy. You have to see attachment style as a property that emerges gradually during thousands of interactions. A child with a particular (genetically influenced) temperament makes bids for protection. A mother with a particular (genetically influenced) temperament responds, or doesn't respond, based on her mood, on how overworked she is, or on what childcare guru she has been reading. No one event is particularly important, but over time the child builds up what Bowlby called an "internal working model" of himself, his mother, and their relationship. If the model says that mom is always there for you, you'll be bolder in your play and explorations. Round after round, predictable and reciprocal interactions build trust and strengthen the relationship. Children with sunny dispositions who have happy mothers are almost certain to play the game well and develop a secure attachment style, but a dedicated mother can overcome either her own or her child's less pleasant disposition and foster a secure internal working model of their relationship. (Everything I have reported above is true for fathers too, but most children in all cultures spend more time with their mothers.)

## IT'S NOT JUST FOR CHILDREN

When I started writing this chapter, I planned to review attachment theory in a page or two and then move on to the stuff that we adults really care about. When we hear the word "love," we think of romantic love. We might hear an occasional song about love between parents and children on a country music radio station, but anywhere else on the dial love means the kind of love you fall into and then struggle to hold onto. The more I delved into the research, however, the more I realized that Harlow, Bowlby, and Ainsworth can help us understand grown-up love. See for yourself. Which of the following statements best describes you in romantic relationships?

1. I find it relatively easy to get close to others and am comfortable depending on them and having them depend on me. I don't often worry about being abandoned or about someone getting too close to me.
2. I am somewhat uncomfortable being close to others; I find it diffi-cult to trust them completely, difficult to allow myself to depend on them. I am nervous when anyone gets too close, and often love part-ners want me to be more intimate than I feel comfortable being.
3. I find that others are reluctant to get as close as I would like. I of-ten worry that my partner doesn't really love me or won't want to stay with me. I want to merge completely with another person, and this desire sometimes scares people away.[20]

The attachment researchers Cindy Hazan and Phil Shaver developed this simple test to see whether Ainsworth's three styles were still at work when adults try to form relationships. They are. Some people change style as they grow up, but the great majority of adults choose the descriptor that matched the way they were as a child.[21] (The three choices above corre-spond to Ainsworth's secure, avoidant, and resistant patterns.) Internal working models are fairly stable (though not unchangeable), guiding people in their most important relationships throughout their lives. And just as secure babies are happier and more well-adjusted, secure adults en-joy happier, longer relationships as well as lower rates of divorce.[22]

But does adult romantic love really grow out of the same psychological system that attaches children to their mothers? To find out, Hazan traced the process by which childhood attachment changes with age. Bowlby had been specific about the four defining features of attachment relationships:[23]

1. proximity maintenance (the child wants and strives to be near the parent)
2. separation distress (self-explanatory)
3. safe haven (the child, when frightened or distressed, comes to the parent for comfort)
4. secure base (the child uses the parent as a base from which to launch exploration and personal growth)

Hazan and her colleagues[24] surveyed hundreds of people from the ages of six through eighty-two, asking which people in their lives fulfilled each of the four defining features of attachment (for example: "Whom do you most like to spend time with?" "Whom do you turn to when you are feeling upset?"). If babies could take the survey, they would nominate mom or dad as the answer to all questions, but by the time they are eight, children want most strongly to spend time with their peers. (When children resist leaving their friends to come home for dinner, that's proximity maintenance.) Between the ages of eight and fourteen, safe haven expands from parents to include peers as adolescents begin turning to each other for emotional support. But it's only at the end of adolescence, around the ages fifteen to seventeen, that all four components of attachment can be satisfied by a peer, specifically a romantic partner. The New Testament records this normal transference of attachment: "For this reason a man shall leave his father and mother and be joined to his wife, and the two shall become one flesh. So they are no longer two, but one flesh" (MARK 10:7–9).

Evidence that romantic partners become true attachment figures, like parents, comes from a review[25] of research on how people cope with the death of a spouse, or a long separation. The review found that adults experience the same sequence Bowlby had observed in children placed in hospitals: initial anxiety and panic, followed by lethargy and depression, followed by recovery through emotional detachment. Furthermore, the review found that contact with close friends was of little help in blunting the pain, but renewed contact with one's *parents* was much more effective.

Once you think about it, the similarities between romantic relationships and parent-infant relationships are obvious. Lovers in the first rush of love spend endless hours in face-to-face mutual gaze, holding each other, nuzzling and cuddling, kissing, using baby voices, and enjoying the same release of the hormone oxytocin that binds mothers and babies to each other in a kind of addiction. Oxytocin prepares female mammals to give birth (triggering uterine contractions and milk release), but it also affects their brains, fostering nurturant behaviors and reducing feelings of stress when mothers are in contact with their children.[26]

This powerful attachment of mothers to infants—often called the "caregiving system"—is a different psychological system from the attachment system in infants, but the two systems obviously evolved in tandem. The infant's distress signals are effective only because they trigger caregiving desires in the mother. Oxytocin is the glue that makes the two parts stick together. Oxytocin has been oversimplified in the popular press as a hormone that makes people (even ornery men) suddenly sweet and affectionate, but more recent work suggests that it can also be thought of as a stress hormone in women.[27] It is secreted when women are under stress and their attachment needs are *not* being met, causing a need for contact with a loved one. On the other hand, when oxytocin floods the brain (male or female) while two people *are* in skin-to-skin contact, the effect is soothing and calming, and it strengthens the bond between them. For adults, the biggest rush of oxytocin—other than giving birth and nursing—comes from sex.[28] Sexual activity, especially if it includes cuddling, extended touching, and orgasm, turns on many of the same circuits that are used to bond infants and parents. It's no wonder that childhood attachment styles persist in adulthood: The whole attachment system persists.

## LOVE AND THE SWELLED HEAD

Adult love relationships are therefore built out of two ancient and interlocking systems: an attachment system that bonds child to mother and a caregiving system that bonds mother to child. These systems are as old as mammals—older perhaps, because birds have them, too. But we still have to add something else to explain why sex is related to love. No problem; nature was motivating animals to seek each other out for sex long before mammals or birds existed. The "mating system" is completely separate from the other two systems, and it involves distinctive brain areas and hormones.[29] In some animals, such as rats, the mating system draws male and female together just long enough for them to copulate. In other species, such as elephants, male and female are drawn together for several days—the duration of the fertile period—during which they share tender caresses, play joyfully, and show many other signs that remind human observers of mutual infatuation.[30] Whatever the duration, for most mammals (other than humans) the three systems are strung together with perfect predictability. First, hormonal changes in the female around the time of ovulation trigger advertisements of her fertility: Female dogs and cats, for example, release pheromones; female chimpanzees and bonobos exhibit enormous red genital swellings. Next, the males become turned on and compete (in some species) to see who gets to mate. The female makes some sort of choice (in most species), which in turn activates her own mating system; and then, some months later, birth activates the caregiving system in the mother and the attachment system in the child. Dad is left out in the cold, where he spends his time sniffing for more pheromones, or scanning for more swellings. Sex is for reproduction; lasting love is for mothers and children. So why are people so different? How did human females come to hide all signs of ovulation and get men to fall in love with them and their children?

Nobody knows, but the most plausible theory[31] in my opinion begins with the enormous expansion of the human brain that I talked about in chapters 1 and 3. When the first hominids split off from the ancestors of modern chimpanzees, their brains were no bigger than those of chimpanzees. These human ancestors were basically just bipedal apes. But then, around 3 million years ago, something changed. Something in the environment, or perhaps an increase in tool use made possible by increasingly dextrous hands, made it highly adaptive to have a much larger brain and much higher intelligence. However, brain growth faced a literal bottleneck: the birth canal. There were physical limits to how large a head hominid females could give birth to and still have a pelvis that would allow them to walk upright. At least one species of hominid—our ancestor—evolved a novel technique that got around this limitation by sending babies out of the uterus long before their brains were developed enough to control their bodies. In all other primate species, brain growth slows dramatically soon after birth because the brain is mostly complete and ready for service; only some fine tuning during a few years of childhood play and learning is needed. In humans, however, the rapid rate of embryonic brain growth continues for about two years after birth, followed by a slower but continuous increase in brain weight for another twenty years.[32] Humans are the only creatures on Earth whose young are utterly helpless for years, and heavily dependent on adult care for more than a decade.

Given the enormous burden that is the human child, women can't do it on their own. Studies of hunter-gatherer societies show that mothers of young children cannot collect enough calories to keep themselves and their children alive.[33] They rely on the large quantity of food as well as the protection provided by males in their peak years of productivity. Big brains, so useful for gossip and social manipulation (as well as hunting and gathering), could therefore have evolved only if men began chipping in. But in the competitive game of evolution, it's a losing move for a male to provide resources to a child who is not his own. So active fathers, male-female pair-bonds, male sexual jealousy, and big-headed babies all co-evolved— that is, arose gradually but together. A man who felt some desire to stay with a woman, guard her fidelity, and contribute to the rearing of their children could produce smarter children than could his less paternal competitors. In environments in which intelligence was highly adaptive (which may have been all human environments, once we began making tools), male investment in children may have paid off for the men themselves (for their genes, that is), and therefore became more common with each successive generation.

But from what raw material could a tie evolve between men and women where one did not exist before? Evolution cannot design anything from scratch. Evolution is a process in which bones and hormones and behavioral patterns that were already coded for by the genes are changed slightly (by random mutation of those genes) and then selected if they confer an advantage on an individual. It didn't take much change to modify the attachment system, which every man and every woman had used as a child to attach to mom, and have it link up with the mating system, which was already turning on in each young person at the time of puberty.

Granted, this theory is speculative (the fossilized bones of a committed father look no different from those of an indifferent one), but it does tie together neatly many of the distinctive features of human life, such as our painful childbirth, long infancy, large brains, and high intelligence. The theory connects these biological quirks about human beings to some of the most important emotional oddities of our species: the existence of strong and (often) enduring emotional bonds between men and women, and between men and children. Because men and women in a relationship have many conflicting interests, evolutionary theory does not view love relationships as harmonious partnerships for childrearing;[34] but a universal feature of human cultures is that men and women form relationships intended to last for years (marriage) that constrain their sexual behavior in some way and institutionalize their ties to children and to each other.

## Two Loves, Two Errors

Take one ancient attachment system, mix with an equal measure of caregiving system, throw in a modified mating system and voila, that's romantic love. I seem to have lost something here; romantic love is so much more than the sum of its parts. It is an extraordinary psychological state that launched the Trojan war, inspired much of the world's best (and worst) music and literature, and gave many of us the most perfect days of our lives. But I think that romantic love is widely misunderstood, and looking at its psychological subcomponents can clear up some puzzles and guide the way around love's pitfalls.

In some corners of universities, the professors tell their students that romantic love is a social construction, invented by the French troubadours of the twelfth century with their stories of chivalry, idealization of women, and the uplifting ache of unconsummated desire. It's certainly true that cultures create their own understandings of psychological phenomena, but many of those phenomena will occur regardless of what people think about them. (For example, death is socially constructed by every culture, but bodies die without consulting those constructions.) A survey of ethnographies from 166 human cultures[35] found clear evidence of romantic love in 88 percent of them; for the rest, the ethnographic record was too thin to be sure either way.

What the troubadours did give us is a particular myth of "true" love—the idea that real love burns brightly and passionately, and then it just keeps on burning until death, and then it just keeps on burning after death as the lovers are reunited in heaven. This myth seems to have grown and diffused in modern times into a set of interrelated ideas about love and marriage. As I see it, the modern myth of true love involves these beliefs: True love is passionate love that never fades; if you are in true love, you should marry that person; if love ends, you should leave that person because it was not true love; and if you can find the right person, you will have true love forever. You might not believe this myth yourself, particularly if you are older than thirty; but many young people in Western nations are raised on it, and it acts as an ideal that they unconsciously carry with them even if they scoff at it. (It's not just Hollywood that perpetrates the myth; Bollywood, the Indian film industry, is even more romanticized.)

But if true love is defined as eternal passion, it is biologically impossible. To see this, and to save the dignity of love, you have to understand the difference between two kinds of love: passionate and companionate. According to the love researchers Ellen Berscheid and Elaine Walster, passionate love is a "wildly emotional state in which tender and sexual feelings, elation and pain, anxiety and relief, altruism and jealousy coexist in a confusion of feelings."[36] Passionate love is the love you fall into. It is what happens when Cupid's golden arrow hits your heart, and, in an instant, the world around you is transformed. You crave union with your beloved. You want, somehow, to crawl into each other. This is the urge that Plato captured in The Symposium, in which Aristophanes' toast to love is a myth about its origins. Aristophanes says that people originally had four legs, four arms, and two faces, but one day the gods felt threatened by the power and arrogance of human beings and decided to cut them in half. Ever since that day, people have wandered the world searching for their other halves. (Some people originally had two male faces, some two female, and the rest a male and a female, thereby explaining the diversity of sexual orientation.) As proof, Aristophanes asks us to imagine that Hephaestus (the god of fire and hence of blacksmiths) were to come upon two lovers as they lay together in an embrace, and say to them:

> What is it you human beings really want from each other? . . . Is this your heart's desire, then—for the two of you to become parts of the same whole, as near as can be, and never to separate, day or night? Because if that's your desire, I'd like to weld you together and join you into something that is naturally whole, so that the two of you are made into one. Then the two of you would share one life, as long as you lived, because you would be one being, and by the same token, when you died, you would be one and not two in Hades, having died a single death. Look at your love, and see if this is what you desire.[37]

Aristophanes says that no lovers would turn down such an offer.

Berscheid and Walster define companionate love, in contrast, as "the affection we feel for those with whom our lives are deeply intertwined."[38] Companionate love grows slowly over the years as lovers apply their attachment and caregiving systems to each other, and as they begin to rely upon, care for, and trust each other. If the metaphor for passionate love is fire, the metaphor for companionate love is vines growing, intertwining, and gradually binding two people together. The contrast of wild and calm forms of love has occurred to people in many cultures. As a woman in a hunter-gatherer tribe in Namibia put it: "When two people come together their hearts are on fire and their passion is very great. After a while, the fire cools and that's how it stays."[39]

Passionate love is a drug. Its symptoms overlap with those of heroin (euphoric well-being, sometimes described in sexual terms) and cocaine (euphoria combined with giddiness and energy).[40] It's no wonder: Passionate love alters the activity of several parts of the brain, including parts that are involved in the release of dopamine.[41] Any experience that feels intensely good releases dopamine, and the dopamine link is crucial here because drugs that artificially raise dopamine levels, as do heroin and cocaine, put you at risk of addiction. If you take cocaine once a month, you won't become addicted, but if you take it every day, you will. No drug can keep you continuously high. The brain reacts to a chronic surplus of dopamine, develops neurochemical reactions that oppose it, and restores its own equilibrium. At that point, tolerance has set in, and when the drug is withdrawn, the brain is unbalanced in the opposite direction: pain, lethargy, and despair follow withdrawal from cocaine or from passionate love.

So if passionate love is a drug—literally a drug—it has to wear off eventually. Nobody can stay high forever (although if you find passionate love in a long-distance relationship, it's like taking cocaine once a month; the drug can retain its potency because of your suffering between doses). If passionate love is allowed to run its joyous course, there must come a day when it weakens. One of the lovers usually feels the change first. It's like waking up from a shared dream to see your sleeping partner drooling. In those moments of returning sanity, the lover may see flaws and defects to which she was blind before. The beloved falls off the pedestal, and then, because our minds are so sensitive to changes, her change in feeling can take on exaggerated importance. "Oh, my God," she thinks, "the magic has worn off— I'm not in love with him anymore." If she subscribes to the myth of true love, she might even consider breaking up with him. After all, if the magic ended, it can't be true love. But if she does end the relationship, she might be making a mistake.

Passionate love does not turn into companionate love. Passionate love and companionate love are two separate processes, and they have different time courses. Their diverging paths produce two danger points, two places where many people make grave mistakes. In figure 6.1, I've drawn out how the intensity of passionate and companionate love might vary in one person's relationship over the course of six months. Passionate love ignites, it burns, and it can reach its maximum temperature within days. During its weeks or months of madness, lovers can't help but think about marriage, and often they talk about it, too. Sometimes they even accept Hephaestus's offer and commit to marriage. This is often a mistake. Nobody can think straight when high on passionate love. The rider is as besotted as the elephant. People are not allowed to sign contracts when they are drunk, and I sometimes wish we could prevent people from proposing marriage when they are high on passionate love because once a marriage proposal is accepted, families are notified, and a date is set, it's very hard to stop the train. The drug is likely to wear off at some point during the stressful wed-

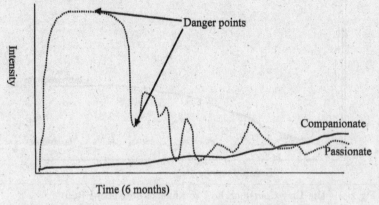

Fig. 6.1  The Time Course of the Two Kinds of Love (Short Run)

ding planning phase, and many of these couples will walk down the aisle with doubt in their hearts and divorce in their future.

The other danger point is the day the drug weakens its grip. Passionate love doesn't end on that day, but the crazy and obsessional high period does. The rider regains his senses and can, for the first time, assess where the elephant has taken them. Breakups often happen at this point, and for many couples that's a good thing. Cupid is usually portrayed as an impish fellow because he's so fond of joining together the most inappropriate couples. But sometimes breaking up is premature, because if the lovers had stuck it out, if they had given companionate love a chance to grow, they might have found true love.

True love exists, I believe, but it is not—cannot be—passion that lasts forever. True love, the love that undergirds strong marriages, is simply strong companionate love, with some added passion, between two people who are firmly committed to each other.[42] Companionate love looks weak in the graph above because it can never attain the intensity of passionate love. But if we change the time scale from six months to sixty years, as in the next figure, it is passionate love that seems trivial—a flash in the pan— while companionate love can last a lifetime. When we admire a couple still in love on their fiftieth anniversary, it is this blend of loves—mostly companionate—that we are admiring.

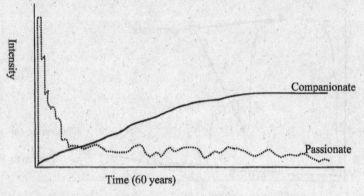

Fig. 6.2   The Time Course of the Two Kinds of Love (Long Run)

## WHY DO PHILOSOPHERS HATE LOVE?

If you are in passionate love and want to celebrate your passion, read poetry. If your ardor has calmed and you want to understand your evolving relationship, read psychology. But if you have just ended a relationship and would like to believe you are better off without love, read philosophy. Oh, there is plenty of work extolling the virtues of love, but when you look closely, you find a deep ambivalence. Love of God, love of neighbor, love of truth, love of beauty—all of these are urged upon us. But the passionate, erotic love of a real person? Heavens no!

In the ancient East, the problem with love is obvious: Love *is* attachment. Attachments, particularly sensual and sexual attachments, must be broken to permit spiritual progress. Buddha said, "So long as lustful desire, however small, of man for women is not controlled, so long the mind of man is not free, but is bound like a calf tied to a cow."[43] *The Laws of Manu,* an ancient Hindu treatise on how young Brahmin men should live, was even more negative about women: "It is the very nature of women to corrupt men here on earth."[44] Even Confucius, who was not focused on breaking attachments, saw romantic love and sexuality as threats to the higher virtues of filial piety and loyalty to one's superiors: "I have never seen anyone who loved virtue as much as sex."[45] (Of course, Buddhism and Hinduism are diverse, and both have changed with time and place. Some modern leaders, such as the Dalai Lama, accept romantic love and its attendant sexuality as an important part of life. But the spirit of the ancient religious and philosophical texts is much more negative.)[46]

In the West, the story is a bit different: Love is widely celebrated by the poets from Homer onwards. Love launches the drama of the *Iliad,* and the *Odyssey* ends with the lusty return of Odysseus to Penelope. When the Greek and Roman philosophers get hold of romantic love, however, they usually either despise it or try to turn it into something else. Plato's *Symposium,* for example, is an entire dialogue devoted to the praise of love. But you never know what position Plato holds until Socrates speaks, and when Socrates speaks, he trashes the eulogies to love that Aristophanes and others have just given. He describes how love produces a "disease" among the animals: "First they are sick for intercourse with each other, then for nurturing their young."[47] (Note: Mating system leads to caregiving system.) For Plato, when human love resembles animal love, it is degrading. The love of a man for a woman, as it aims at procreation, is therefore a debased kind of love. Plato's Socrates then shows how love can transcend its animal origins by aiming at something higher. When an older man loves a young man, their love can be elevating for both because the older man can, in between rounds of intercourse, teach the young man about virtue and philosophy. But even this love must be a stepping stone only: When a man loves a beautiful body he must learn to love beauty in general, not the beauty of one particular body. He must come to find beauty in men's souls, and then in ideas and philosophy. Ultimately he comes to know the form of beauty itself:

> The result is that he will see the beauty of knowledge and be looking mainly not at beauty in a single example—as a servant would who favored the beauty of a little boy or a man or a single custom . . . but the lover is turned to the great sea of beauty, and, gazing upon this, he gives birth to many gloriously beautiful ideas and theories, in unstinting love of wisdom. . . .[48]

The essential nature of love as an attachment between two people is rejected; love can be dignified only when it is converted into an appreciation of beauty in general.

The later Stoics also object to the particularity of love, to the way it places the source of one's happiness in the hands of another person, whom one cannot fully control. Even the Epicureans, whose philosophy was based on the pursuit of pleasure, value friendship but oppose romantic love. In *De Rerum Natura*, the philosophical poet Lucretius lays out the fullest surviving statement of the philosophy of Epicurus. The end of Book 4 is widely known as the "Tirade Against Love," in which Lucretius compares love to a wound, a cancer, and a sickness. The Epicureans were experts on desire and its satisfaction; they objected to passionate love because it cannot be satisfied:

> When two lie tasting, limb by limb
> life's bloom, when flesh gives foretaste of delight,
> and Venus is ready to sow the female field,
> they hungrily seize each other, mouth to mouth
> the spittle flows, they pant, press tooth to lip—
> vainly, for they can chafe no substance off
> nor pierce and be gone, one body in the other.
> For often this seems to be their wish, their goal,
> so greedily do they cling in passion's bond.[49]

Christianity brought forward many of these classical fears of love. Jesus commands his followers to love God, using the same words as Moses ("With all your heart, and with all your soul, and with all your might," MATTHEW, 22:37, in referring to DEUTERONOMY 6:5). Jesus' second commandment is to love one another: "You shall love your neighbor as yourself" (MATTHEW 22:39). But what can it mean to love others as one loves oneself? The psychological origins of love are in attachment to parents and sexual partners. We do not attach to ourselves; we do not seek security and fulfillment in ourselves. What Jesus seems to mean is that we should *value* others as much as we value ourselves; we should be kind and generous even to strangers and even to our enemies. This uplifting message is relevant to the issues of reciprocity and hypocrisy that I talked about in chapters 3 and 4, but it has little to do with the psychological systems I have been covering in this chapter. Rather, Christian love has focused on two key words: *caritas* and *agape*. *Caritas* (the origin of our word "charity") is a kind of intense benevolence and good will; *agape* is a Greek word that refers to a kind of selfless, spiritual love with no sexuality, no clinging to a particular other person. (Of course, Christianity endorses the love of a man and a woman within marriage, but even this love is idealized as the love of Christ for his church—EPHESIANS 5:25) As in Plato, Christian love is love stripped of its essential particularity, its focus on a *specific* other person. Love is remodeled into a general attitude toward a much larger, even infinite, class of objects.

Caritas and agape are beautiful, but they are not related to or derived from the kinds of love that people *need*. Although I would like to live in a world in which everyone radiates benevolence toward everyone else, I would rather live in a world in which there was at least one person who loved me specifically, and whom I loved in return. Suppose Harlow had raised rhesus monkeys under two conditions. For the first group, each was reared in its own cage, but each day Harlow put in a new but very nurturing adult female monkey as a companion. For the second group, each was reared in a cage with its own mother, and then each day Harlow put in a new and not particularly nice other monkey. The monkeys in the first group got something like caritas—benevolence without particularity—and they would probably emerge emotionally damaged. Without having formed an attachment relationship, they would likely be fearful of new experiences and unable to love or care for other monkeys. The monkeys in the second group would have had something closer to a normal rhesus monkey childhood, and would probably emerge healthy and able to love. Monkeys and people need close and long-lasting attachments to particular others. In chapter 9, I will propose that agape is real, but usually short-lived. It can change lives and enrich lives, but it cannot substitute for the kinds of love based on attachments.

There are several reasons why real human love might make philosophers uncomfortable. First, passionate love is notorious for making people illogical and irrational, and Western philosophers have long thought that morality is grounded in rationality. (In chapter 8, I will argue against this view.) Love is a kind of insanity, and many people have, while crazed with passion, ruined their lives and those of others. Much of the philosophical opposition to love may therefore be well-intentioned advice by the sages to the young: Shut your ears to the sirens' deceitful song.

I think, however, that at least two less benevolent motivations are at work. First, there may be a kind of hypocritical self-interest in which the older generation says, "Do as we say, not as we did." Buddha and St. Augustine, for example, drank their fill of passionate love as young men and came out only much later as opponents of sexual attachments. Moral codes are designed to keep order within society; they urge us to rein in our desires and play our assigned roles. Romantic love is notorious for making young people give less than a damn about the rules and conventions of their society, about caste lines, or about feuds between Capulets and Montagues. So the sages' constant attempts to redefine love as something spiritual and prosocial sound to me like the moralism of parents who, having enjoyed a variety of love affairs when they were young, now try to explain to their daughter why she should save herself for marriage.

A second motivation is the fear of death. Jamie Goldenberg[50] at the University of Colorado has shown that when people are asked to reflect on their own mortality, they find the physical aspects of sexuality more disgusting, and they are less likely to agree with an essay arguing for the essential similarity of people and animals. Goldenberg and her colleagues believe that people in all cultures have a pervasive fear of death. Human beings all know that they are going do die, and so human cultures go to great lengths to construct systems of meaning that dignify life and convince people that their lives have more meaning than those of the animals that die all around them. The extensive regulation of sex in many cultures, the attempt to link love to God and then to cut away the sex, is part of an elaborate defense against the gnawing fear of mortality.[51]

If this is true, if the sages have a variety of unstated reasons for warning us away from passionate love and attachments of many kinds, perhaps we should be selective in heeding their advice. Perhaps we need to look at our own lives, lived in a world very different from theirs, and also at the evidence about whether attachments are good or bad for us.

## FREEDOM CAN BE
## HAZARDOUS TO YOUR HEALTH

In the late nineteenth century, one of the founders of sociology, Emile Durkheim, performed a scholarly miracle. He gathered data from across

Europe to study the factors that affect the suicide rate. His findings can be summarized in one word: constraints. No matter how he parsed the data, people who had fewer social constraints, bonds, and obligations were more likely to kill themselves. Durkheim looked at the "degree of integration of religious society" and found that Protestants, who lived the least demanding religious lives at the time, had higher suicide rates than did Catholics; Jews, with the densest network of social and religious obligations, had the lowest. He examined the "degree of integration of domestic society"—the family—and found the same thing: People living alone were most likely to kill themselves; married people, less; married people with children, still less. Durkheim concluded that people need obligations and constraints to provide structure and meaning to their lives: "The more weakened the groups to which [a man] belongs, the less he depends on them, the more he consequently depends only on himself and recognizes no other rules of conduct than what are founded on his private interests."[52]

A hundred years of further studies have confirmed Durkheim's diagnosis. If you want to predict how happy someone is, or how long she will live (and if you are not allowed to ask about her genes or personality), you should find out about her social relationships. Having strong social relationships strengthens the immune system, extends life (more than does quitting smoking), speeds recovery from surgery, and reduces the risks of depression and anxiety disorders.[53] It's not just that extroverts are naturally happier and healthier; when introverts are forced to be more outgoing, they usually enjoy it and find that it boosts their mood.[54] Even people who think they don't want a lot of social contact still benefit from it. And it's not just that "we all need somebody to lean on"; recent work on *giving* support shows that caring for others is often more beneficial than is receiving help.[55] We need to interact and intertwine with others; we need the give *and* the take; we need to belong.[56] An ideology of extreme personal freedom can be dangerous because it encourages people to leave homes, jobs, cities, and marriages in search of personal and professional fulfillment, thereby breaking the relationships that were probably their best hope for such fulfillment.

Seneca was right: "No one can live happily who has regard to himself alone and transforms everything into a question of his own utility." John

Donne was right: No man, woman, or child is an island. Aristophanes was right: We need others to complete us. We are an ultrasocial species, full of emotions finely tuned for loving, befriending, helping, sharing, and otherwise intertwining our lives with others. Attachments and relationships can bring us pain: As a character in Jean-Paul Sartre's play *No Exit* said, "Hell is other people."[57] But so is heaven.

# 7

## The Uses of Adversity

*When heaven is about to confer a great responsibility on any man, it will exercise his mind with suffering, subject his sinews and bones to hard work, expose his body to hunger, put him to poverty, place obstacles in the paths of his deeds, so as to stimulate his mind, harden his nature, and improve wherever he is incompetent.*

— MENG TZU,[1] CHINA, 3RD CENT. BCE

*What doesn't kill me makes me stronger.*

— NIETSZCHE[2]

MANY TRADITIONS HAVE a notion of fate, predestination, or divine fore-knowledge. Hindus have a folk belief that on the day of birth, God writes the destiny of each child upon his or her forehead. Suppose that on the day your child is born, you are given two gifts: a pair of glasses that allows you to read this forecast, and a pencil that allows you to edit it. (Suppose further that the gifts come from God, with full permission to use them as you please.) What would you do? You read the list: At age nine: best friend dies of cancer. At eighteen: graduates high school at top of class. At twenty: car accident while driving drunk leads to amputation of left leg. At twenty-four: becomes single parent. At twenty-nine: marries. At thirty-two: publishes successful novel. At thirty-three: divorces; and so on. How painful you'd find it to see your child's future suffering written out before you! What parent could resist the urge to cross off the traumas, to correct the self-inflicted wounds?

But be careful with that pencil. Your good intentions could make things worse. If Nietzsche is right that what doesn't kill you makes you stronger, then the complete erasure of serious adversity from your child's future would leave him or her weak and underdeveloped. This chapter is about what we might call the "adversity hypothesis," which says that people need adversity, setbacks, and perhaps even trauma to reach the highest levels of strength, fulfillment, and personal development.

Nietzsche's dictum can't be literally true, at least, not all the time. People who face the real and present threat of their own deaths, or who witness the violent deaths of others, sometimes develop posttraumatic stress disorder (PTSD), a debilitating condition that leaves its victims anxious and over-reactive. People who suffer from PTSD are changed, sometimes perma-nently: They panic or crumble more easily when faced with later adversity. Even if we take Nietzsche figuratively (which he would have much pre-ferred anyway), fifty years of research on stress shows that stressors are gen-erally bad for people,[3] contributing to depression, anxiety disorders, and heart disease. So let's be cautious about accepting the adversity hypothesis. Let's look to scientific research to figure out when adversity is beneficial, and when it is harmful. The answer is not just "adversity within limits." It's a much more interesting story, one that reveals how human beings grow and thrive, and how you (and your child) can best profit from the adversity that surely lies in your future.

## POSTTRAUMATIC GROWTH

Greg's life fell apart on April 8, 1999. On that day, his wife and two chil-dren, ages four and seven, disappeared. It took Greg three days just to find out that they had not died in a car crash; Amy had taken the children and run off with a man she had met in a shopping mall a few weeks earlier. The four of them were now driving around the country and had been spotted in several Western states. The private detective Greg hired quickly discovered that the man who had ruined Greg's life earned his living as a con artist and petty criminal. How could this have happened? Greg felt like Job, stripped in one day of all he loved most. And like Job, he had no explanation for what had befallen him.

Greg,[4] an old friend of mine, called me to see whether I, as a psychologist, could offer insight into how his wife had fallen under the influence of such a fraud. The one insight I could offer was that the man sounded like a psychopath. Most psychopaths are not violent (although most serial murderers and serial rapists are psychopaths). They are people, mostly men, who have no moral emotions, no attachment systems, and no concerns for others.[5] Because they feel no shame, embarrassment, or guilt, they find it easy to manipulate people into giving them money, sex, and trust. I told Greg that if this man was indeed a psychopath, he was incapable of love and would soon tire of Amy and the kids. Greg would probably see his children again soon.

Two months later, Amy returned. The police restored the children to Greg's custody. Greg's panic phase was over, but so was his marriage, and Greg began the long and painful process of rebuilding his life. He was now a single parent living on an assistant professor's salary, and he faced years of legal expenses fighting Amy over the custody of their children. He had little hope of finishing the book his academic career depended upon, and he worried about his children's mental health, and his own. What was he going to do?

I visited Greg a few months later. It was a beautiful August evening, and as we sat on his porch, Greg told me about how the crisis had affected him. He was still in pain, but he had learned that many people cared about him and were there to help him. Families from his church were bringing him meals and helping out with childcare. His parents were selling their house in Utah and moving to Charlottesville to help him raise the children. Also, Greg said that the experience had radically changed his perspective about what mattered in life. As long as he had his children back, career success was no longer so important to him. Greg said he now treated people differently, a change related to his change in values: He found himself reacting to others with much greater sympathy, love, and forgiveness. He just couldn't get mad at people for little things anymore. And then Greg said something so powerful that I choked up. Referring to the often sad and moving solo that is at the heart of many operas, he said: "This is my moment to sing the aria. I don't want to, I don't want to have this chance, but it's here now, and what am I going to do about it? Am I going to rise to the occasion?"

To have framed things in such a way showed that he was already rising. With the help of family, friends, and deep religious faith, Greg rebuilt his life, finished his book, and two years later found a better job. When I spoke to him recently, he told me he still feels wounded by what happened. But he also said that many of the positive changes had endured, and that he now experiences more joy from each day with his children than he did before the crisis.

For decades, research in health psychology focused on stress and its damaging effects. A major concern in this research literature has always been resilience—the ways people cope with adversity, fend off damage, and "bounce back" to normal functioning. But it's only in the last fifteen years that researchers have gone beyond resilience and begun to focus on the *benefits* of severe stress. These benefits are sometimes referred to collectively as "posttraumatic growth,"[6] in direct contrast to posttraumatic stress disorder. Researchers have now studied people facing many kinds of adversity, including cancer, heart disease, HIV, rape, assault, paralysis, infertility, house fires, plane crashes, and earthquakes. Researchers have studied how people cope with the loss of their strongest attachments: children, spouses or partners, and parents. This large body of research shows that although traumas, crises, and tragedies come in a thousand forms, people benefit from them in three primary ways—the same ones that Greg talked about.

The first benefit is that rising to a challenge reveals your hidden abilities, and seeing these abilities changes your self-concept. None of us knows what we are really capable of enduring. You might say to yourself, "I would die if I lost X," or "I could never survive what Y is going through," yet these are statements spun out of thin air by the rider. If you did lose X, or find yourself in the same position as Y, your heart would not stop beating. You would respond to the world as you found it, and most of those responses would be automatic. People sometimes say they are numb or on autopilot after a terrible loss or trauma. Consciousness is severely altered, yet somehow the body keeps moving. Over the next few weeks some degree of normalcy returns as one struggles to make sense of the loss and of one's altered circumstances. What doesn't kill you makes you, by definition, a survivor, about whom people then say, "I could never survive what Y is going through." One of the most common lessons people draw from bereavement or trauma is that they are much stronger than they realized, and this new appreciation of their strength then gives them confidence to face future challenges. And they are not just confabulating a silver lining to wrap around a dark cloud; people who have suffered through battle, rape, concentration camps, or traumatic personal losses often seem to be inoculated[7] against future stress: They recover more quickly, in part because they know they can cope. Religious leaders have often pointed to exactly this benefit of suffering. As Paul said in his Letter to the Romans (5:3–4): "Suffering produces endurance, and endurance produces character, and character produces hope." More recently, the Dalai Lama said: "The person who has had more experience of hardships can stand more firmly in the face of problems than the person who has never experienced suffering. From this angle, then, some suffering can be a good lesson for life."[8]

The second class of benefit concerns relationships. Adversity is a filter. When a person is diagnosed with cancer, or a couple loses a child, some friends and family members rise to the occasion and look for any way they can to express support or to be helpful. Others turn away, perhaps unsure of what to say or unable to overcome their own discomfort with the situation. But adversity doesn't just separate the fair-weather friends from the true; it strengthens relationships and it opens people's hearts to one another. We often develop love for those we care for, and we usually feel love and gratitude toward those who cared for us in a time of need. In a large study of bereavement, Susan Nolen-Hoeksema and her colleagues at Stanford University found that one of the most common effects of losing a loved one was that the bereaved had a greater appreciation of and tolerance for the other people in his or her life. A woman in the study, whose partner had died of cancer, explained: "[The loss] enhanced my relationship with other people because I realize that time is so important, and you can waste so much effort on small, insignificant events or feelings."[9] Like Greg, this bereaved woman found herself relating to others in a more loving and less petty way. Trauma seems to shut off the motivation to play Machiavellian tit for tat with its emphasis on self-promotion and competition.

This change in ways of relating points to the third common benefit: Trauma changes priorities and philosophies toward the present ("Live each day to the fullest") and toward other people. We have all heard stories about rich and powerful people who had a moral conversion when faced with death. In 1993, I saw one of the grandest such stories written in the rocks outside the Indian city of Bhubaneswar, where I spent three months studying culture and morality. King Ashoka, after assuming control of the Maurya empire (in central India) around 272 BCE, set out to expand his territory by conquest. He was successful, subduing by slaughter many of the peoples and kingdoms around him. But after a particularly bloody victory over the Kalinga people, near what is now Bhubaneswar, he was seized with horror and remorse. He converted to Buddhism, renounced all further conquest by violence, and devoted his life to creating a kingdom based on justice and respect for dharma (the cosmic law of Hinduism and Buddhism). He wrote out his vision of a just society and his rules for virtuous behavior, and had these edicts carved into rock walls throughout his kingdom. He sent emissaries as far away as Greece to spread his vision of peace, virtue, and religious tolerance. Ashoka's conversion was caused by victory, not adversity, yet people are often traumatized—as modern research on soldiers[10] indicates—by killing as well as by facing the threat of death. Like so many who experience posttraumatic growth, Ashoka underwent a profound transformation. In his edicts, he described himself as having become more forgiving, compassionate, and tolerant of those who differed with him.

Few people have the chance to go from mass murderer to patron of humanity, but a great many people facing death report changes in values and perspectives. A diagnosis of cancer is often described, in retrospect, as a wake-up call, a reality check, or a turning point. Many people consider changing careers or reducing the time they spend at work. The reality that people often wake up to is that life is a gift they have been taking for granted, and that people matter more than money. Charles Dickens's *A Christmas Carol* captures a deep truth about the effects of facing mortality: A few minutes with the ghost of "Christmas Yet to Come" converts Scrooge, the ultimate miser, into a generous man who takes delight in his family, his employees, and the strangers he passes on the street.

I don't want to celebrate suffering, prescribe it for everyone, or minimize the moral imperative to reduce it where we can. I don't want to ignore the pain that ripples out from each diagnosis of cancer, spreading fear along lines of kinship and friendship. I want only to make the point that suffering is not always all bad for all people. There is usually some good mixed in with the bad, and those who find it have found something precious: a key to moral and spiritual development. As Shakespeare wrote:

> Sweet are the uses of adversity,
> Which like the toad, ugly and venomous,
> Wears yet a precious jewel in his head.[11]

## MUST WE SUFFER?

The adversity hypothesis has a weak and a strong version. In the weak version, adversity *can* lead to growth, strength, joy, and self-improvement, by the three mechanisms of posttraumatic growth described above. The weak version is well-supported by research, but it has few clear implications for how we should live our lives. The strong version of the hypothesis is more unsettling: It states that people *must* endure adversity to grow, and that the highest levels of growth and development are *only* open to those who have faced and overcome great adversity. If the strong version of the hypothesis is valid, it has profound implications for how we should live our lives and structure our societies. It means that we should take more chances and suffer more defeats. It means that we might be dangerously overprotecting our children, offering them lives of bland safety and too much counseling while depriving them of the "critical incidents"[12] that would help them to grow strong and to develop the most intense friendships. It means that heroic societies, which fear dishonor more than death, or societies that struggle together through war, might produce better human beings than can a world of peace and prosperity in which people's expectations rise so high that they sue each other for "emotional damages."

But is the strong version valid? People often say that they have been profoundly changed by adversity, yet researchers have so far collected little evidence of adversity-induced personality change beyond such reports.
People's scores on personality tests are fairly stable over the course of a few
years, even for people who report that they have changed a great deal in
the interim.[13] In one of the few studies that tried to verify reports of growth
by asking the subjects' friends about them, the friends noticed much less
change than the subjects had reported.[14]

These studies might, however, have been looking for change in the wrong
place. Psychologists often approach personality by measuring basic traits
such as the "big five": neuroticism, extroversion, openness to new experiences, agreeableness (warmth/niceness), and conscientiousness.[15] These
traits are facts about the elephant, about a person's automatic reactions to
various situations. They are fairly similar between identical twins reared
apart, indicating that they are influenced in part by genes, although they are
also influenced by changes in the conditions of one's life or the roles one
plays, such as becoming a parent.[16] But psychologist Dan McAdams has
suggested that personality really has three levels,[17] and too much attention
has been paid to the lowest level, the basic traits. A second level of personality, "characteristic adaptations," includes personal goals, defense and coping
mechanisms, values, beliefs, and life-stage concerns (such as those of parent-
hood or retirement) that people develop to succeed in their particular roles
and niches. These adaptations are influenced by basic traits: A person high
on neuroticism will have many more defense mechanisms; an extrovert will
rely more heavily on social relationships. But in this middle level, the person's
basic traits are made to mesh with facts about the person's environment and
stage of life. When those facts change—as after losing a spouse—the person's characteristic adaptations change. The elephant might be slow to
change, but the elephant and rider, working together, find new ways of getting through the day.

The third level of personality is that of the "life story." Human beings in
every culture are fascinated by stories; we create them wherever we can.
(See those seven stars up there? They are seven sisters who once . . . ) It's
no different with our own lives. We can't stop ourselves from creating what

McAdams describes as an "evolving story that integrates a reconstructed past, perceived present, and anticipated future into a coherent and vitalizing life myth."[18] Although the lowest level of personality is mostly about the elephant, the life story is written primarily by the rider. You create your story in consciousness as you interpret your own behavior, and as you listen to other people's thoughts about you. The life story is not the work of a historian—remember that the rider has no access to the *real* causes of your behavior; it is more like a work of historical fiction that makes plenty of references to real events and connects them by dramatizations and interpretations that might or might not be true to the spirit of what happened.

From this three-level perspective, it becomes clear why adversity might be necessary for optimal human development. Most of the life goals that people pursue at the level of "characteristic adaptations" can be sorted—as the psychologist Robert Emmons[19] has found—into four categories: work and achievement, relationships and intimacy, religion and spirituality, and generativity (leaving a legacy and contributing something to society). Although it is generally good for you to pursue goals, not all goals are equal. People who strive primarily for achievement and wealth are, Emmons finds, less happy, on average, than those whose strivings focus on the other three categories.[20] The reason takes us back to happiness traps and conspicuous consumption (see chapter 5): Because human beings were shaped by evolutionary processes to pursue success, not happiness, people enthusiastically pursue goals that will help them win prestige in zero-sum competitions. Success in these competitions feels good but gives no lasting pleasure, and it raises the bar for future success.

When tragedy strikes, however, it knocks you off the treadmill and forces a decision: Hop back on and return to business as usual, or try something else? There is a window of time—just a few weeks or months after the tragedy—during which you are more open to something else. During this time, achievement goals often lose their allure, sometimes coming to seem pointless. If you shift toward other goals—family, religion, or helping others—you shift to inconspicuous consumption, and the pleasures derived along the way are not fully subject to adaptation (treadmill) effects. The pursuit of these goals therefore leads to more happiness but less wealth (on average). Many people change their goals in the wake of adversity; they resolve to work less, to love and play more. If in those first few months you take action—you do something that changes your daily life—then the changes might stick. But if you do nothing more than make a resolution ("I must never forget my new outlook on life"), then you will soon slip back into old habits and pursue old goals. The rider can exert some influence at forks in the road; but the elephant handles daily life, responding automatically to the environment. Adversity may be necessary for growth because it forces you to stop speeding along the road of life, allowing you to notice the paths that were branching off all along, and to think about where you really want to end up.

At the third level of personality, the need for adversity is even more obvious: You need interesting material to write a good story. McAdams says that stories are "fundamentally about the vicissitudes of human intention organized in time."[21] You can't have a good life story without vicissitudes, and if the best you can come up with is that your parents refused to buy you a sports car for your sixteenth birthday, nobody will want to read your memoirs. In the thousands of life stories McAdams has gathered, several genres are associated with well-being. For example, in the "commitment story," the protagonist has a supportive family background, is sensitized early in life to the sufferings of others, is guided by a clear and compelling personal ideology, and, at some point, transforms or redeems failures, mistakes, or crises into a positive outcome, a process that often involves setting new goals that commit the self to helping others. The life of the Buddha is a classic example.

In contrast, some people's life stories show a "contamination" sequence in which emotionally positive events go bad and everything is spoiled. People who tell such stories are, not surprisingly, more likely to be depressed.[22] Indeed, part of the pathology of depression is that, while ruminating, the depressed person reworks her life narrative by using the tools of Beck's negative triad: I'm bad, the world is bad, and my future is dark. Although adversity that is not overcome can create a story of depressing bleakness, substantial adversity might be necessary for a meaningful story.

McAdams's ideas are profoundly important for understanding posttraumatic growth. His three levels of personality allow us to think about *coher-*

*ence* among the levels. What happens when the three levels of personality don't match up? Imagine a woman whose basic traits are warm and gregarious but who strives for success in a career that offers few chances for close contacts with people, and whose life story is about an artist forced by her parents to pursue a practical career. She is a mess of mismatched motives and stories, and it may be that only through adversity will she be able to make the radical changes she would need to achieve coherence among levels. The psychologists Ken Sheldon and Tim Kasser have found that people who are mentally healthy and happy have a higher degree of "vertical coherence" among their goals—that is, higher-level (long term) goals and lower-level (immediate) goals all fit together well so that pursuing one's short-term goals advances the pursuit of long-term goals.[23]

Trauma often shatters belief systems and robs people of their sense of meaning. In so doing, it forces people to put the pieces back together, and often they do so by using God or some other higher purpose as a unifying principle.[24] London and Chicago seized the opportunities provided by their great fires to remake themselves into grander and more coherent cities. People sometimes seize such opportunities, too, rebuilding beautifully those parts of their lives and life stories that they could never have torn down voluntarily. When people report having grown after coping with adversity, they could be trying to describe a new sense of inner coherence. This coherence might not be visible to one's friends, but it feels like growth, strength, maturity, and wisdom from the inside.[25]

## BLESSED ARE THE SENSE MAKERS

When bad things happen to good people, we have a problem. We know consciously that life is unfair, but unconsciously we see the world through the lens of reciprocity. The downfall of an evil man (in our biased and moralistic assessment) is no puzzle: He had it coming to him. But when the victim was virtuous, we struggle to make sense of his tragedy. At an intuitive level, we all believe in karma, the Hindu notion that people reap what they sow. The psychologist Mel Lerner has demonstrated that we are so motivated to believe that people get what they deserve and deserve what they get that we often blame the victim of a tragedy, particularly when we can't achieve justice by punishing a perpetrator or compensating the victim.[26]

In Lerner's experiments, the desperate need to make sense of events can lead people to inaccurate conclusions (for example, a woman "led on" a rapist); but, in general, the ability to make sense of tragedy and then find benefit in it is the key that unlocks posttraumatic growth.[27] When trauma strikes, some people find the key dangling around their necks with instructions printed on it. Others are left to fend for themselves, and they do not fend as well. Psychologists have devoted a great deal of effort to figuring out who benefits from trauma and who is crushed. The answer compounds the already great unfairness of life: Optimists are more likely to benefit than pessimists.[28] Optimists are, for the most part, people who won the cortical lottery: They have a high happiness setpoint, they habitually look on the bright side, and they easily find silver linings. Life has a way of making the rich get richer and the happy get happier.

When a crisis strikes, people cope in three primary ways:[29] active coping (taking direct action to fix the problem), reappraisal (doing the work within—getting one's own thoughts right and looking for silver linings), and avoidance coping (working to blunt one's emotional reactions by denying or avoiding the events, or by drinking, drugs, and other distractions). People who have a basic-level trait of optimism (McAdams's level 1) tend to develop a coping style (McAdams's level 2) that alternates between active coping and reappraisal. Because optimists expect their efforts to pay off, they go right to work fixing the problem. But if they fail, they expect that things usually work out for the best, and so they can't help but look for possible benefits. When they find them, they write a new chapter in their life story (McAdams's level 3), a story of continual overcoming and growth. In contrast, people who have a relatively negative affective style (complete with more activity in the front right cortex than the front left) live in a world filled with many more threats and have less confidence that they can deal with them. They develop a coping style that relies more heavily on avoidance and other defense mechanisms. They work harder to manage their pain than to fix their problems, so their problems often get worse. Drawing the lesson that the world is unjust and uncontrollable, and that things often work out for the worst, they weave this lesson into their life story where it contaminates the narrative.

If you are a pessimist, you are probably feeling gloomy right now. But despair not! The key to growth is not optimism per se; it is the sense making that optimists find easy. If you can find a way to make sense of adversity and draw constructive lessons from it, you can benefit, too. And you can learn to become a sense maker by reading Jamie Pennebaker's *Opening Up*.[30] Pennebaker began his research by studying the relationship between trauma, such as childhood sexual abuse, and later health problems. Trauma and stress are usually bad for people, and Pennebaker thought that self-disclosure—talking with friends or therapists—might help the body at the same time that it helps the mind. One of his early hypotheses was that traumas that carry more shame, such as being raped (as opposed to a nonsexual assault) or losing a spouse to suicide (rather than to a car accident), would produce more illness because people are less likely to talk about such events with others. But the nature of the trauma turned out to be almost irrelevant. What mattered was what people did afterward: Those who talked with their friends or with a support group were largely spared the health-damaging effects of trauma.

Once Pennebaker had found a correlation between disclosure and health, he took the next step in the scientific process and tried to *create* health benefits by getting people to disclose their secrets. Pennebaker asked people to write about "the most upsetting or traumatic experience of your entire life," preferably one they had not talked about with others in great detail. He gave them plenty of blank paper and asked them to keep writing for fifteen minutes, on four consecutive days. Subjects in a control group were asked to write about some other topic (for example, their houses, a typical work day) for the same amount of time. In each of his studies, Pennebaker got his subjects' permission to obtain their medical records at some point in the future. Then he waited a year and observed how often people in the two groups got sick. The people who wrote about traumas went to the doctor or the hospital fewer times in the following year. I did not believe this result when I first heard it. How on earth could one hour of writing stave off the flu six months later? Pennebaker's results seemed to support an old-fashioned Freudian notion of catharsis: People who express their emotions, "get it off their chests" or "let off steam," are healthier. Having once reviewed the literature on the catharsis hypothesis, I knew that there was no evidence for it.[31] Letting off steam makes people angrier, not calmer.

Pennebaker discovered that it's not about steam; it's about sense making. The people in his studies who used their writing time to vent got no benefit. The people who showed deep insight into the causes and consequences of the event on their first day of writing got no benefit, either: They had already made sense of things. It was the people who made *progress* across the four days, who showed increasing insight; they were the ones whose health improved over the next year. In later studies, Pennebaker asked people to dance or sing to express their emotions, but these emotionally expressive activities gave no health benefit.[32] You have to use words, and the words have to help you create a meaningful story. If you can write such a story you can reap the benefits of reappraisal (one of the two healthy coping styles) even years after an event. You can close a chapter of your life that was still open, still affecting your thoughts and preventing you from moving on with the larger narrative.

Anyone, therefore, can benefit from adversity, although a pessimist will have to take some extra steps, some conscious, rider-initiated steps, to guide the elephant gently in the right direction. The first step is to do what you can, before adversity strikes, to change your cognitive style. If you are a pessimist, consider meditation, cognitive therapy, or even Prozac. All three will make you less subject to negative rumination, more able to guide your thoughts in a positive direction, and therefore more able to withstand future adversity, find meaning in it, and grow from it. The second step is to cherish and build your social support network. Having one or two good attachment relationships helps adults as well as children (and rhesus monkeys) to face threats. Trusted friends who are good listeners can be a great aid to making sense and finding meaning. Third, religious faith and practice can aid growth, both by directly fostering sense making (religions provide stories and interpretive schemes for losses and crises) and by increasing social support (religious people have relationships through their religious communities, and many have a relationship with God). A portion of the benefits of religiosity[33] could also be a result of the confession and disclosure of inner turmoil, either to God or to a religious authority that many religions encourage.

And finally, no matter how well or poorly prepared you are when trouble strikes, at some point in the months afterwards, pull out a piece of paper and start writing. Pennebaker suggests[34] that you write continuously for fifteen minutes a day, for several days. Don't edit or censor yourself; don't worry about grammar or sentence structure; just keep writing. Write about what happened, how you feel about it, and *why* you feel that way. If you hate to write, you can talk into a tape recorder. The crucial thing is to get your thoughts and feelings out without imposing any order on them—but in such a way that, after a few days, some order is likely to emerge on its own. Before you conclude your last session, be sure you have done your best to answer these two questions: Why did this happen? What good might I derive from it?

## FOR EVERYTHING THERE IS A SEASON

If the adversity hypothesis is true, and if the mechanism of benefit has to do with sense making and getting those three levels of personality to cohere, then there should be times in life when adversity will be more or less beneficial. Perhaps the strong version of the hypothesis is true during only a part of the life course?

There are many reasons for thinking that children are particularly vulnerable to adversity. Genes guide brain development throughout childhood, but that development is also affected by environmental context, and one of the most important contextual factors is the overall level of safety versus threat. Good parenting can help tune up the attachment system to make a child more adventurous; yet, even beyond such effects, if a child's environment feels safe and controllable, the child will (on average) develop a more positive affective style, and will be less anxious as an adult.[35] But if the environment offers daily uncontrollable threats (from predators, bullies, or random violence), the child's brain will be altered, set to be less trusting and more vigilant.[36] Given that most people in modern Western nations live in safe worlds where optimism and approach motivations generally pay off, and given that most people in psychotherapy need loosening up, not tightening up, it is probably best for children to develop the most positive affective style, or the highest set range (S from chapter 5), that their genes will allow. Major adversity is unlikely to have many—or perhaps any—beneficial effects for children. (On the other hand, children are amazingly resilient and are not as easily damaged by one-time events, even by sexual abuse, as most people think.[37] Chronic conditions are much more important.) Of course, children need limits to learn self-control, and they need plenty of failure to learn that success takes hard work and persistence. Children should be protected, but not spoiled.

Things might be different for teenagers. Younger children know some stories about themselves, but the active and chronic striving to integrate one's past, present, and future into a coherent narrative begins only in the mid to late teens.[38] This claim is supported by a curious fact about autobiographical memory called the "memory bump." When people older than thirty are asked to remember the most important or vivid events of their lives, they are disproportionately likely to recall events that occurred between the ages of fifteen and twenty-five.[39] This is the age when a person's life blooms—first love, college and intellectual growth, living and perhaps traveling independently—and it is the time when young people (at least in Western countries) make many of the choices that will define their lives. If there is a special period for identify formation, a time when life events are going to have the biggest influence on the rest of the life-story, this is it. So adversity, especially if overcome fully, is probably most beneficial in the late teens and early twenties.

We can't ethically conduct experiments that induce trauma at different ages, but in a way life has performed these experiments for us. The major events of the twentieth century—the Great Depression, World War II—hit people at different ages, and the sociologist Glen Elder[40] has produced elegant analyses of longitudinal data (collected from the same people over many decades) to find out why some thrived and others crumbled after these adversities hit. Elder once summarized his findings this way: "There is a storyline across all the work I've done. Events do not have meaning in themselves. Those meanings are derived from the interactions between people, groups, and the experience itself. Kids who went through very difficult circumstances usually came out rather well."[41] Elder found that a lot hinged on the family and the person's degree of social integration: Children as well as adults who weathered crises while embedded within strong social groups and networks fared much better; they were more likely to come out stronger and mentally healthier than were those who faced adversity without such social support. Social networks didn't just reduce suffering, they offered avenues for finding meaning and purpose (as Durkheim concluded from his studies of suicide).[42] For example, the widely shared adversity of the Great Depression offered many young people the chance to make a real contribution to their families by finding a job that brought in a few dollars a week. The need for people to pull together within their nations to fight World War II appears to have made those who lived through it more responsible and civic minded, at least in the United States, even if they played no direct role in the war effort.[43]

There is, however, a time limit on first adversity. Elder says that life starts to "crystallize" by the late twenties. Even young men who had not been doing well before serving in World War II often turned their lives around afterward, but people who faced their first real life test after the age of thirty (for example, combat in that war, or financial ruin in the Great Depression) were less resilient and less likely to grow from their experiences. So adversity may be most beneficial for people in their late teens and into their twenties.

Elder's work is full of reminders that the action is in the interactions—that is, the ways that one's unique personality interacts with details about an event and its social context to produce a particular and often unpredictable outcome. In the area of research known as "life-span development,"[44] there are few simple rules in the form of "X causes Y." Nobody, therefore, can propose an ideal life course with carefully scheduled adversity that would be beneficial for everyone. We can say, however, that for many people, particularly those who overcame adversity in their twenties, adversity made them stronger, better, and even happier than they would have been without it.

## ERROR AND WISDOM

I expect that when I have children, I'll be no different from other parents in wanting to edit their forehead writing and remove all adversity. Even if I could be convinced that a trauma experienced at the age of twenty-four was going to teach my daughter important lessons and make her a better person, I'd think: Well, why can't I just teach her those lessons directly? Isn't there some way she can reap the benefits without the costs? But a common piece of worldly wisdom is that life's most important lessons cannot be taught directly. Marcel Proust said:

> We do not receive wisdom, we must discover it for ourselves, after a jour-
> ney through the wilderness which no one else can make for us, which no
> one can spare us, for our wisdom is the point of view from which we
> come at last to regard the world.[45]

Recent research on wisdom proves Proust correct. Knowledge comes in two major forms: explicit and tacit. Explicit knowledge is all the facts you know and can consciously report, independent of context. Wherever I am, I know that the capital of Bulgaria is Sofia. Explicit knowledge is taught directly in schools. The rider gathers it up and files it away, ready for use in later reasoning. But wisdom is based—according to Robert Sternberg,[46] a leading wisdom researcher—on "tacit knowledge." Tacit knowledge is procedural (it's "knowing how" rather than "knowing that"), it is acquired without direct help from others, and it is related to goals that a person values. Tacit knowledge resides in the elephant. It's the skills that the elephant acquires, gradually, from life experience. It depends on context: There is no universal set of best practices for ending a romantic relationship, consoling a friend, or resolving a moral disagreement.

Wisdom, says Sternberg, is the tacit knowledge that lets a person balance two sets of things. First, wise people are able to balance their own needs, the needs of others, and the needs of people or things beyond the immediate interaction (e.g., institutions, the environment, or people who may be adversely affected later on). Ignorant people see everything in black and white—they rely heavily on the myth of pure evil—and they are strongly influenced by their own self-interest. The wise are able to see things from others' points of view, appreciate shades of gray, and then choose or advise a course of action that works out best for everyone in the long run. Second, wise people are able to balance three responses to situations: adaptation (changing the self to fit the environment), shaping (changing the environment), and selection (choosing to move to a new environment). This second balance corresponds roughly to the famous "serenity prayer": "God, grant me the serenity to accept the things I cannot change, courage to change the things I can, and wisdom to know the difference."[47] If you already know this prayer, your rider knows it (explicitly). If you live this prayer, your elephant knows it, too (tacitly), and you are wise.

Sternberg's ideas show why parents can't teach their children wisdom directly. The best they can do is provide a range of life experiences that will help their children acquire tacit knowledge in a variety of life domains. Parents can also model wisdom in their own lives and gently encourage children to think about situations, look at other viewpoints, and achieve balance in challenging times. Shelter your children when young, but if the sheltering goes on through the child's teens and twenties, it may keep out wisdom and growth as well as pain. Suffering often makes people more compassionate, helping them find balance between self and others. Suffering often leads to active coping (Sternberg's shaping), reappraisal coping (Sternberg's adaptation), or changes in plans and directions (Sternberg's selection). Posttraumatic growth usually involves, therefore, the growth of wisdom.

The strong version of the adversity hypothesis might be true, but only if we add caveats: For adversity to be maximally beneficial, it should happen at the right time (young adulthood), to the right people (those with the social and psychological resources to rise to challenges and find benefits), and to the right degree (not so severe as to cause PTSD). Each life course is so unpredictable that we can never know whether a particular setback will be beneficial to a particular person in the long run. But perhaps we do know enough to allow some editing of a child's forehead writing: Go ahead and erase some of those early traumas, but think twice, or await future research, before erasing the rest.

# 8

❖

# The Felicity of Virtue

> *It is impossible to live the pleasant life without also living*
> *sensibly, nobly and justly, and it is impossible to live sensibly,*
> *nobly and justly without living pleasantly.*
>
> —EPICURUS[1]

> *Set your heart on doing good. Do it over and over again, and*
> *you will be filled with joy. A fool is happy until his mischief*
> *turns against him. And a good man may suffer until his good-*
> *ness flowers.*
>
> —BUDDHA[2]

WHEN SAGES AND ELDERS urge virtue on the young, they sometimes sound like snake-oil salesmen. The wisdom literature of many cultures essentially says, "Gather round, I have a tonic that will make you happy, healthy, wealthy, and wise! It will get you into heaven, and bring you joy on earth along the way! Just be virtuous!" Young people are extremely good, though, at rolling their eyes and shutting their ears. Their interests and desires are often at odds with those of adults; they quickly find ways to pursue their goals and get themselves into trouble, which often becomes character-building adventure. Huck Finn runs away from his foster mother to raft down the Mississippi with an escaped slave; the young Buddha

155

leaves his father's palace to begin his spiritual quest in the forest; Luke Skywalker leaves his home planet to join the galactic rebellion. All three set off on epic journeys that make each into an adult, complete with a set of new virtues. These hard-won virtues are especially admirable to us as readers because they reveal a depth and authenticity of character that we don't see in the obedient kid who simply accepts the virtues he was raised with.

In this light, Ben Franklin is supremely admirable. Born in Boston in 1706, he was apprenticed at the age of twelve to his older brother James, who owned a printing shop. After many disputes with (and beatings from) his brother, he yearned for freedom, but James would not release him from the legal contract of his apprenticeship. So at the age of seventeen, Ben broke the law and skipped town. He got on a boat to New York and, failing to find work there, kept on going to Philadelphia. There he found work as an apprentice printer and, through skill and diligence, eventually opened his own print shop and published his own newspaper. He went on to spectacular success in business (*Poor Richard's Almanack*—a compendium of sayings and maxims—was a hit in its day); in science (he proved that lightning is electricity, then tamed it by inventing the lightning rod); in politics (he held too many offices to name); and in diplomacy (he persuaded France to join the American colonies' war against Britain, though France had little to gain from the enterprise). He lived to eighty-four and enjoyed the ride. He took pride in his scientific discoveries and civic creations; he basked in the love and esteem of France as well as of America; and even as an old man he relished the attentions of women.

What was his secret? Virtue. Not the sort of uptight, pleasure-hating Puritanism that some people now associate with that word, but a broader kind of virtue that goes back to ancient Greece. The Greek word *aretē* meant excellence, virtue, or goodness, especially of a functional sort. The *aretē* of a knife is to cut well; the *aretē* of an eye is to see well; the *aretē* of a person is . . . well, that's one of the oldest questions of philosophy: What is the true nature, function, or goal of a person, relative to which we can say that he or she is living well or badly? Thus in saying that well being or happiness (*eudaimonia*) is "an activity of soul in conformity with excellence or virtue,"[3] Aristotle wasn't saying that happiness comes from giving to the poor and suppressing your sexuality. He was saying that a good life is one where you develop your strengths, realize your potential, and become what it is in your nature to become. (Aristotle believed that all things in the universe had a *telos*, or purpose toward which they aimed, even though he did not believe that the gods had designed all things.)

One of Franklin's many gifts was his extraordinary ability to see potential and then realize it. He saw the potential of paved and lighted streets, volunteer fire departments, and public libraries, and he pushed to make them all appear in Philadelphia. He saw the potential of the young American republic and played many roles in creating it. He also saw the potential in himself for improving his ways, and he set out to do so. In his late twenties, as a young printer and entrepreneur, he embarked on what he called a "bold and arduous project of arriving at moral perfection."[4] He picked a few virtues he wanted to cultivate, and he tried to live accordingly. He discovered immediately the limitations of the rider:

> While my care was employed in guarding against one fault, I was often surprised by another; habit took the advantage of inattention; inclination was sometimes too strong for reason. I concluded, at length, that the mere speculative conviction that it was our interest to be completely virtuous was not sufficient to prevent our slipping, and that the contrary habits must be broken, and good ones acquired and established, before we can have any dependence on a steady, uniform rectitude of conduct.[5]

Franklin was a brilliant intuitive psychologist. He realized that the rider can be successful only to the extent that it trains the elephant (though he did not use those terms), so he devised a training regimen. He wrote out a list of thirteen virtues, each linked to specific behaviors that he should or should not do. (For example: "Temperance: Eat not to dullness"; "Frugality: Make no expense but to do good to others or yourself"; "Chastity: Rarely use venery but for health or offspring"). He then printed a table made up of seven columns (one for each day of the week) and thirteen rows (one for each virtue), and he put a black spot in the appropriate square each time he failed to live a whole day in accordance with a particular virtue. He concentrated on only one virtue a week, hoping to keep its row clear of spots while paying no special attention to the other virtues, though he filled in their rows whenever violations occurred. Over thirteen weeks, he worked through the whole table. Then he repeated the process, finding that with repetition the table got less and less spotty. Franklin wrote in his autobiography that, though he fell far short of perfection: "I was, by the endeavor, a better and a happier man than I otherwise should have been if I had not attempted it." He went on: "My posterity should be informed that to this little artifice, with the blessing of God, their ancestor ow'd the constant felicity of his life, down to his 79th year, in which this is written."[6]

We can't know whether, without his virtue table, Franklin would have been any less happy or successful, but we can search for other evidence to test his main psychological claim. This claim, which I will call the "virtue hypothesis," is the same claim made by Epicurus and the Buddha in the epigraphs that open this chapter: Cultivating virtue will make you happy. There are plenty of reasons to doubt the virtue hypothesis. Franklin himself admitted that he failed utterly to develop the virtue of humility, yet he reaped great social gains by learning to fake it. Perhaps the virtue hypothesis will turn out to be true only in a cynical, Machiavellian way: Cultivating the *appearance* of virtue will make you successful, and therefore happy, regardless of your true character.

## THE VIRTUES OF THE ANCIENTS

Ideas have pedigrees, ideas have baggage. When we Westerners think about morality, we use concepts that are thousands of years old, but that took a turn in their development in the last two hundred years. We don't realize that our approach to morality is odd from the perspective of other cultures, or that it is based on a particular set of psychological assumptions—a set that now appears to be wrong.

Every culture is concerned about the moral development of its children, and in every culture that left us more than a few pages of writing, we find texts that reveal its approach to morality. Specific rules and prohibitions vary, but the broad outlines of these approaches have a lot in common. Most cultures wrote about virtues that should be cultivated, and many of those virtues were and still are valued across most cultures[7] (for example, honesty, justice, courage, benevolence, self-restraint, and respect for authority). Most approaches then specified actions that were good and bad with respect to those virtues. Most approaches were practical, striving to inculcate virtues that would benefit the person who cultivates them.

One of the oldest works of direct moral instruction is the *Teaching of Amenemope*, an Egyptian text thought to have been written around 1300 BCE. It begins by describing itself as "instruction about life" and as a "guide for well-being," promising that whoever commits its lessons to heart will "discover . . . a treasure house of life, and [his] body will flourish upon earth." Amenemope then offers thirty chapters of advice about how to treat other people, develop self-restraint, and find success and contentment in the process. For example, after repeatedly urging honesty, particularly in respecting the boundary markers of other farmers, the text says:

> Plow your fields, and you'll find what you need,
> You'll receive bread from your threshing floor.
> Better is a bushel given you by God
> Than five thousand through wrongdoing. . . .
> Better is bread with a happy heart
> Than wealth with vexation.[8]

If this last line sounds familiar to you, it is because the biblical book of Proverbs borrowed a lot from Amenemope. For example: "Better is a little with the fear of the Lord than great treasure and trouble with it" (PROVERBS 15:16).

An additional common feature is that these ancient texts rely heavily on maxims and role models rather than proofs and logic. Maxims are carefully phrased to produce a flash of insight and approval. Role models are presented to elicit admiration and awe. When moral instruction triggers emotions, it speaks to the elephant as well as the rider. The wisdom of Confucius and Buddha, for example, comes down to us as lists of aphorisms so timeless and evocative that people still read them today for pleasure and guidance, refer to them as "worldwide laws of life,"[9] and write books about their scientific validity.

A third feature of many ancient texts is that they emphasize practice and habit rather than factual knowledge. Confucius compared moral development to learning how to play music;[10] both require the study of texts, observance of role models, and many years of practice to develop "virtuosity." Aristotle used a similar metaphor:

> Men become builders by building houses, and harpists by playing the harp. Similarly, we grow just by the practice of just actions, self-controlled by exercising our self-control, and courageous by performing acts of courage.[11]

Buddha offered his followers the "Eightfold Noble Path," a set of activities that will, with practice, create an ethical person (by right speech, right action, right livelihood), and a mentally disciplined person (by right effort, right mindfulness, right concentration).

In all these ways, the ancients reveal a sophisticated understanding of moral psychology, similar to Franklin's. They all knew that virtue resides in a well-trained elephant. They all knew that training takes daily practice and a great deal of repetition. The rider must take part in the training, but if moral instruction imparts only *explicit* knowledge (facts that the rider can state), it will have no effect on the elephant, and therefore little effect on behavior. Moral education must also impart *tacit* knowledge—skills of social perception and social emotion so finely tuned that one automatically *feels* the right thing in each situation, *knows* the right thing to do, and then *wants* to do it. Morality, for the ancients, was a kind of practical wisdom.

## HOW THE WEST WAS LOST

The Western approach to morality got off to a great start; as in other ancient cultures, it focused on virtues. The Old Testament, the New Testament, Homer, and Aesop all show that our founding cultures relied heavily on proverbs, maxims, fables, and role models to illustrate and teach the virtues. Plato's *Republic* and Aristotle's *Nichomachean Ethics*, two of the greatest works of Greek philosophy, are essentially treatises on the virtues and their cultivation. Even the Epicureans, who thought pleasure was the goal of life, believed that people needed virtues to cultivate pleasures.

Yet contained in these early triumphs of Greek philosophy are the seeds of later failure. First, the Greek mind that gave us moral inquiry also gave us the beginnings of scientific inquiry, the aim of which is to search for the smallest set of laws that can explain the enormous variety of events in the world. Science values parsimony, but virtue theories, with their long lists of virtues, were never parsimonious. How much more satisfying it would be to the scientific mind to have one virtue, principle, or rule from which all others could be derived? Second, the widespread philosophical worship of reason made many philosophers uncomfortable with locating virtue in habits and feelings. Although Plato located most of virtue in the rationality of his charioteer, even he had to concede that virtue required the right passions; he therefore came up with that complicated metaphor in which one of two horses contains some virtue, but the other has none. For Plato and many later thinkers, rationality was a gift from the gods, a tool to control our animal lusts. Rationality had to be in charge.

These two seeds—the quest for parsimony and the worship of reason— lay dormant in the centuries after the fall of Rome, but they sprouted and bloomed in the European Enlightenment of the eighteenth century. As advances in technology and commerce began to create a new world, some people began to seek rationally justified social and political arrangements. The French philosopher René Descartes, writing in the seventeenth century, was quite happy to rest his ethical system on the benevolence of God, but Enlightenment thinkers sought a foundation for ethics that did not depend on divine revelation or on God's enforcement. It was as though somebody had offered a prize, like the prizes that lured early aviators to undertake daring journeys: Ten thousand pounds sterling to the first philosopher who can come up with a single moral rule, to be applied through the power of reason, that can cleanly separate good from bad.

Had there been such a prize, it would have gone to the German philosopher Immanuel Kant.[12] Like Plato, Kant believed that human beings have a dual nature: part animal and part rational. The animal part of us follows the laws of nature, just as does a falling rock or a lion killing its prey. There is no morality in nature; there is only causality. But the rational part of us, Kant said, can follow a different kind of law: It can respect rules of conduct, and so people (but not lions) can be judged morally for the degree to which they respect the right rules. What might those rules be? Here Kant devised the cleverest trick in all moral philosophy. He reasoned that for moral rules to be *laws*, they had to be universally applicable. If gravity worked differently for men and women, or for Italians and Egyptians, we could not speak of it as a law. But rather than searching for rules to which all people would in fact agree (a difficult task, likely to produce only a few bland generalities), Kant turned the problem around and said that people should think about whether the rules guiding their own actions could reasonably be *proposed* as universal laws. If you are planning to break a promise that has become inconvenient, can you really propose a universal rule that states people *ought* to break promises that have become inconvenient? Endorsing such a rule would render all promises meaningless. Nor could you consistently will that people cheat, lie, steal, or in any other way deprive other people of their rights or their property, for such evils would surely come back to visit you. This simple test, which Kant called the "categorical imperative," was extraordinarily powerful. It offered to make ethics a branch of applied logic, thereby giving it the sort of certainty that secular ethics, without recourse to a sacred book, had always found elusive.

Over the following decades, the English philosopher Jeremy Bentham challenged Kant for the (hypothetical) prize. When Bentham became a lawyer in 1767, he was appalled by the complexities and inefficiencies of English law. He set out, with typical enlightenment boldness, to re-conceive the entire legal and legislative system by stating clear goals and proposing the most rational means of achieving those goals. The ultimate goal of all legislation, he concluded, was the good of the people; and the more good, the better. Bentham was the father of utilitarianism, the doctrine that in all decisionmaking (legal and personal), our goal should be the maximum total benefit (utility), but who gets the benefit is of little concern.[13]

The argument between Kant and Bentham has continued ever since. Descendants of Kant (known as "deontologists" from the Greek *deon*, obligation) try to elaborate the duties and obligations that ethical people must respect, even when their actions lead to bad outcomes (for example, you must never kill an innocent person, even if doing so will save a hundred lives). Descendants of Bentham (known as "consequentialists" because they evaluate actions only by their consequences) try to work out the rules and policies that will bring about the greatest good, even when doing so will sometimes violate other ethical principles (go ahead and kill the one to save the hundred, they say, unless it will set a bad example that leads to later problems).

Despite their many differences, however, the two camps agree in important ways. They both believe in parsimony: Decisions should be based ultimately on one principle only, be it the categorical imperative or the maximization of utility. They both insist that only the rider can make such decisions because moral decision making requires logical reasoning and sometimes even mathematical calculation. They both distrust intuitions and gut feelings, which they see as obstacles to good reasoning. And they both shun the particular in favor of the abstract: You don't need a rich, thick description of the people involved, or of their beliefs and cultural traditions. You just need a few facts and a ranked list of their likes and dislikes (if you are a utilitarian). It doesn't matter what country or historical era you are in; it doesn't matter whether the people involved are your friends, your enemies, or complete strangers. The moral law, like a law of physics, works the same for all people at all times.

These two philosophical approaches have made enormous contributions to legal and political theory and practice; indeed, they helped create societies that respect individual rights (Kant) while still working efficiently for the good of the people (Bentham). But these ideas have also permeated Western culture more generally, where they have had some unintended consequences. The philosopher Edmund Pincoffs[14] has argued that consequentialists and deontologists worked together to convince Westerners in the twentieth century that morality is the study of moral quandaries and dilemmas. Where the Greeks focused on the *character* of a person and asked what kind of person we should each aim to become, modern ethics focuses on *actions,* asking when a particular action is right or wrong. Philosophers wrestle with life-and-death dilemmas: Kill one to save five? Allow aborted fetuses to be used as a source of stem cells? Remove the feeding tube from a woman who has been unconscious for fifteen years? Nonphilosophers wrestle with smaller quandaries: Pay my taxes when others are cheating? Turn in a wallet full of money that appears to belong to a drug dealer? Tell my spouse about a sexual indiscretion?

This turn from character ethics to quandary ethics has turned moral education away from virtues and toward moral reasoning. If morality is about dilemmas, then moral education is training in problem solving. Children must be taught how to think about moral problems, especially how to overcome their natural egoism and take into their calculations the needs of others. As the United States became more ethnically diverse in the 1970s and 1980s, and also more averse to authoritarian methods of education, the idea of teaching specific moral facts and values went out of fashion. Instead, the rationalist legacy of quandary ethics gave us teachers and many parents who would enthusiastically endorse this line, from a recent child-rearing handbook: "My approach does not teach children what and what not to do and why, but rather, it teaches them how to think so they can decide for themselves what and what not to do, and why."[15]

I believe that this turn from character to quandary was a profound mistake, for two reasons. First, it weakens morality and limits its scope. Where the ancients saw virtue and character at work in everything a person does, our modern conception confines morality to a set of situations that arise for each person only a few times in any given week: tradeoffs between self-interest and the interests of others. In our thin and restricted modern conception, a moral person is one who gives to charity, helps others, plays by the rules, and in general does not put her own self-interest too far ahead of others'. Most of the activities and decisions of life are therefore insulated from moral concern. When morality is reduced to the opposite of self-interest, however, the virtue hypothesis becomes paradoxical: In modern terms, the virtue hypothesis says that acting against your self-interest is in your self-interest. It's hard to convince people that this is true, and it can't possibly be true in all situations. In his time, Ben Franklin had a much easier task when he extolled the virtue hypothesis. Like the ancients, he had a thicker, richer notion of virtues as a garden of excellences that a person cultivates to become more effective and appealing to others. Seen in this way, virtue is, obviously, its own reward. Franklin's example implicitly posed this question for his contemporaries and his descendants: Are you willing to work now for your own later well-being, or are you so lazy and short-sighted that you won't make the effort?

The second problem with the turn to moral reasoning is that it relies on bad psychology. Many moral education efforts since the 1970s take the rider off of the elephant and train him to solve problems on his own. After being exposed to hours of case studies, classroom discussions about moral dilemmas, and videos about people who faced dilemmas and made the right choices, the child learns how (not what) to think. Then class ends, the rider gets back on the elephant, and nothing changes at recess. Trying to make children behave ethically by teaching them to reason well is like trying to make a dog happy by wagging its tail. It gets causality backwards.

During my first year of graduate school at the University of Pennsylvania, I discovered the weakness of moral reasoning in myself. I read a wonderful book—*Practical Ethics*—by the Princeton philosopher Peter Singer.[16] Singer, a humane consequentialist, shows how we can apply a consistent concern for the welfare of others to resolve many ethical problems of daily life. Singer's approach to the ethics of killing animals changed forever my thinking about my food choices. Singer proposes and justifies a few guiding principles: First, it is wrong to cause pain and suffering to any sentient creature, therefore current factory farming methods are unethical. Second, it is wrong to take the life of a sentient being that has some sense of identity and attachments, therefore killing animals with large brains and highly developed social lives (such as other primates and most other mammals) is wrong, even if they could be raised in an environment they enjoyed and were then killed painlessly. Singer's clear and compelling arguments convinced me on the spot, and since that day I have been morally opposed to all forms of factory farming. Morally opposed, but not behaviorally opposed. I love the taste of meat, and the only thing that changed in the first six months after reading Singer is that I thought about my hypocrisy each time I ordered a hamburger.

But then, during my second year of graduate school, I began to study the emotion of disgust, and I worked with Paul Rozin, one of the foremost authorities on the psychology of eating. Rozin and I were trying to find video clips to elicit disgust in the experiments we were planning, and we met one morning with a research assistant who showed us some videos he had found. One of them was *Faces of Death*, a compilation of real and fake video footage of people being killed. (These scenes were so disturbing that we could not ethically use them.) Along with the videotaped suicides and executions, there was a long sequence shot inside a slaughterhouse. I watched in horror as cows, moving down a dripping disassembly line, were bludgeoned, hooked, and sliced up. Afterwards, Rozin and I went to lunch to talk about the project. We both ordered vegetarian meals. For days afterwards, the sight of red meat made me queasy. My visceral feelings now matched the beliefs Singer had given me. The elephant now agreed with the rider, and I became a vegetarian. For about three weeks. Gradually, as the disgust faded, fish and chicken reentered my diet. Then red meat did, too, although even now, eighteen years later, I still eat less red meat and choose non-factory-farmed meats when they are available.

That experience taught me an important lesson. I think of myself as a fairly rational person. I found Singer's arguments persuasive. But, to paraphrase Medea's lament (from chapter 1): I saw the right way and approved it, but followed the wrong, until an emotion came along to provide some force.

## THE VIRTUES OF POSITIVE PSYCHOLOGY

The cry that we've lost our way is heard from some quarter in every country and era, but it has been particularly loud in the United States since the social turmoil of the 1960s and the economic malaise and rising crime of the 1970s. Political conservatives, particularly those who have strong religious beliefs, bridled at the "value-free" approach to moral education and the "empowering" of children to think for themselves instead of teaching them facts and values to think about. In the 1980s, these conservatives challenged the education establishment by pushing for character education programs in schools, and by home-schooling their own children.

Also in the 1980s, several philosophers helped to revive virtue theories. Most notably, Alasdair MacIntyre argued in *After Virtue*[17] that the "enlightenment project" of creating a universal, context-free morality was doomed from the beginning. Cultures that have shared values and rich traditions invariably generate a framework in which people can value and evaluate each other. One can easily talk about the virtues of a priest, a soldier, a mother, or a merchant in the context of fourth-century BCE Athens. Strip away all identity and context, however, and there is little to grab on to. How much can you say about the virtues of a generalized *Homo sapiens,* floating in space with no particular sex, age, occupation, or culture? The modern requirement that ethics ignore particularity is what gave us our weaker morality—applicable everywhere, but encompassing nowhere. MacIntyre says that the loss of a language of virtue, grounded in a particular tradition, makes it difficult for us to find meaning, coherence, and purpose in life.[18]

In recent years, even psychology has become involved. In 1998, Martin Seligman founded positive psychology when he asserted that psychology had lost its way. Psychology had become obsessed with pathology and the dark side of human nature, blind to all that was good and noble in people. Seligman noted that psychologists had created an enormous manual, known as the "DSM" (the *Diagnostic and Statistical Manual of Mental Disorders*), to diagnose every possible mental illness and behavioral annoyance, but psychology didn't even have a language with which to talk about the upper reaches of human health, talent, and possibility. When Seligman launched positive psychology, one of his first goals was to create a diagnostic manual for the strengths and virtues. He and another psychologist, Chris Peterson of the University of Michigan, set out to construct a list of the strengths and virtues, one that might be valid for any human culture. I argued with them that the list did *not* have to be valid for all cultures to be useful; they should focus just on large-scale industrial societies. Several anthropologists told them that a universal list could never be created. Fortunately, however, they persevered.

As a first step, Peterson and Seligman surveyed every list of virtues they could find, from the holy books of major religions down to the Boy Scout Oath ("trustworthy, loyal, helpful, friendly . . . "). They made large tables of virtues and tried to see which ones were common across lists. Although no specific virtue made every list, six broad virtues, or families of related virtues, appeared on nearly all lists: wisdom, courage, humanity, justice, temperance, and transcendence (the ability to forge connections to something larger than the self). These virtues are widely endorsed because they are abstract: There are many ways to be wise, or courageous, or humane, and it is impossible to find a human culture that rejects all forms of any of these virtues. (Can we even imagine a culture in which parents hope that their children will grow up to be foolish, cowardly, and cruel?) But the real value of the list of six is that it serves as an organizing framework for more specific *strengths of character*. Peterson and Seligman define character strengths as specific ways of displaying, practicing, and cultivating the virtues. Several paths lead to each virtue. People, as well as cultures, vary in the degree to which they value each path. This is the real power of the classification: It points to specific means of growth toward widely valued ends without insisting that any one way is mandatory for all people at all times. The classification is a tool for diagnosing people's diverse strengths and for helping them find ways to cultivate excellence.

Peterson and Seligman suggest that there are twenty-four principle character strengths, each leading to one of the six higher-level virtues.[19] You can diagnose yourself by looking at the list below or by taking the strengths test (at www.authentichappiness.org).

1. Wisdom:
   - Curiosity
   - Love of learning
   - Judgment
   - Ingenuity
   - Emotional intelligence
   - Perspective
2. Courage:
   - Valor
   - Perseverance
   - Integrity
3. Humanity:
   - Kindness
   - Loving
4. Justice:
   - Citizenship
   - Fairness
   - Leadership

5. Temperance:
   - Self-control
   - Prudence
   - Humility
6. Transcendence:
   - Appreciation of beauty and excellence
   - Gratitude
   - Hope
   - Spirituality
   - Forgiveness
   - Humor
   - Zest

Odds are that you don't have much trouble with the list of six virtue families, but you do have objections to the longer list of strengths. Why is humor a means to transcendence? Why is leadership on the list, but not the virtues of followers and subordinates—duty, respect, and obedience? Please, go ahead and argue. The genius of Peterson and Seligman's classification is to get the conversation going, to propose a specific list of strengths and virtues, and then let the scientific and therapeutic communities work out the details. Just as the DSM is thoroughly revised every ten or fifteen years, the classification of strengths and virtues (known among positive psychologists as the "un-DSM") is sure to be revised and improved in a few years. In daring to be specific, in daring to be wrong, Peterson and Seligman have demonstrated ingenuity, leadership, and hope.

This classification is already generating exciting research and liberating ideas. Here's my favorite idea: Work on your strengths, not your weaknesses. How many of your New Year's resolutions have been about fixing a flaw? And how many of those resolutions have you made several years in a row? It's difficult to change any aspect of your personality by sheer force of will, and if it is a weakness you choose to work on, you probably won't enjoy the process. If you don't find pleasure or reinforcement along the way, then—unless you have the willpower of Ben Franklin—you'll soon give up. But you don't really have to be good at everything. Life offers so many chances to use one tool instead of another, and often you can use a strength to get around a weakness.

In the positive psychology class I teach at the University of Virginia, the final project is to make yourself a better person, using all the tools of psychology, and then prove that you have done so. About half the students each year succeed, and the most successful ones usually either use cognitive behavioral therapy on themselves (it really does work!) or employ a strength, or both. For example, one student lamented her inability to forgive. Her mental life was dominated by ruminations about how those to whom she was closest had hurt her. For her project, she drew on her strength of loving: Each time she found herself spiraling down into thoughts about victimhood, she brought to mind a positive memory about the person in question, which triggered a flash of affection. Each flash cut off her anger and freed her, temporarily, from rumination. In time, this effortful mental process became habitual and she became more forgiving (as she demonstrated using the reports she had filled out each day to chart her progress). The rider had trained the elephant with rewards at each step.

Another outstanding project was done by a woman who had just undergone surgery for brain cancer. At the age of twenty-one, Julia faced no better than even odds of surviving. To deal with her fears, she cultivated one of her strengths—zest. She made lists of the activities going on at the university and of the beautiful hikes and parks in the nearby Blue Ridge Mountains. She shared these lists with the rest of the class, she took time away from her studies to go on these hikes, and she invited friends and classmates to join her. People often say that adversity makes them want to live each day to the fullest, and when Julia made a conscious effort to cultivate her natural strength of zest, she really did it. (She is still full of zest today.)

Virtue sounds like hard work, and often is. But when virtues are reconceived as excellences, each of which can be achieved by the practice of several strengths of character, and when the practice of these strengths is often intrinsically rewarding, suddenly the work sounds more like Csik-szentmihalyi's flow and less like toil. It's work that—like Seligman's description of gratifications—engages you fully, draws on your strengths, and allows you to lose self-consciousness and immerse yourself in what you are doing. Franklin would be pleased: The virtue hypothesis is alive and well, firmly ensconced in positive psychology.

## HARD QUESTION, EASY ANSWERS

Virtue can be its own reward, but that's obvious only for the virtues that one finds rewarding. If your strengths include curiosity or love of learning, you'll enjoy cultivating wisdom by traveling, going to museums, and attending public lectures. If your strengths include gratitude and appreciation of beauty, the feelings of transcendence you get from contemplating the Grand Canyon will give you pleasure too. But it would be naive to think that doing the right thing always feels good. The real test of the virtue hypothesis is to see whether it is true even in our restricted modern understanding of morality as altruism. Forget all that stuff about growth and excellence. Is it true that acting against my self-interest, for the good of others, even when I don't want to, is still good for me? Sages and moralists have always answered with an unqualified yes, but the challenge for science is to qualify: When is it true, and why?

Religion and science each begin with an easy and unsatisfying answer, but then move on to more subtle and interesting explanations. For religious sages, the easy way out is to invoke divine reciprocity in the afterlife. Do good, because God will punish the wicked and reward the virtuous. For Christians, there's heaven or hell. Hindus have the impersonal workings of karma: The universe will repay you in the next life with a higher or lower rebirth, which will depend upon your virtue in this life.

I'm in no position to say whether God, heaven, or an afterlife exists, but as a psychologist I am entitled to point out that belief in postmortem justice shows two signs of primitive moral thinking. In the 1920s, the great developmental psychologist Jean Piaget[20] got down on his knees to play marbles and jacks with children and, in the process, mapped out how morality develops. He found that, as children develop an increasingly sophisticated understanding of right and wrong, they go through a phase in which many rules take on a kind of sacredness and unchangeability. During this phase, children believe in "immanent justice"—justice that is inherent in an act itself. In this stage, they think that if they break rules, even accidentally, something bad will happen to them, even if nobody knows about their transgressions. Immanent justice shows up in adults, too, particularly when it comes to explaining illness and grave misfortune. A survey[21] of beliefs about the causes of illness across cultures shows that the three most common explanations are biomedical (referring to physical causes of disease), interpersonal (illness is caused by witchcraft, related to envy and conflict), and moral (illness is caused by one's own past actions, particularly violations of food and sexual taboos). Most Westerners consciously embrace the biomedical explanation and reject the other two, yet when illness strikes and Westerners ask, "Why me?" one of the places they often look for answers is to their own past transgressions. The belief that God or fate will dole out rewards and punishments for good and bad behavior seems on its face to be a cosmic extension of our childhood belief in immanent justice, which is itself a part of our obsession with reciprocity.

The second problem with postmortem justice is that it relies on the myth of pure evil.[22] Each of us can easily divide the world into good and evil, but presumably God would not suffer from the many biases and Machiavellian motivations that make us do so. Moral motivations (justice, honor, loyalty, patriotism) enter into most acts of violence, including terrorism and war. Most people believe their actions are morally justified. A few paragons of evil stand out as candidates for hell, but almost everyone else would end up in limbo. It just won't work to turn God into Santa Claus, a moral accountant keeping track of 6 billion accounts, because most lives can't be placed definitively in the naughty or nice columns.

The scientific approach to the question also begins with an easy and unsatisfying answer: Virtue is good for your genes under some circumstances. When "survival of the fittest" came to mean "survival of the fittest gene," it became easy to see that the fittest genes would motivate kind and cooperative behavior in two scenarios: when it benefited those who bore a copy of those genes (that is, kin), or when it benefited the bearers of the genes directly by helping them reap the surplus of non-zero-sum games using the tit-for-tat strategy. These two processes—kin altruism and reciprocal altruism—do indeed explain nearly all altruism among nonhuman animals, and much of human altruism, too. This answer is unsatisfying, however, because our genes are, to some extent, puppet masters making us want things that are sometimes good for them but bad for us (such as extramarital affairs, or prestige bought at the expense of happiness). We cannot look to genetic self-interest as a guide either to virtuous or to happy living. Furthermore, anyone who does embrace reciprocal altruism as a *justification* for altruism (rather than merely a cause of it) would then be free to pick and choose: Be nice to those who can help you, but don't waste time or money on anyone else (for example, never leave a tip in restaurants you will not return to). So to evaluate the idea that altruism pays for the altruist, we need to push the sages and the scientists harder: Does it even pay when there is neither postmortem nor reciprocal payback?

## HARD QUESTION, HARD ANSWERS

St. Paul quotes Jesus as having said that "it is more blessed to give than to receive" (ACTS 20:35). One meaning of "bless" is "to confer happiness or prosperity upon."[23] Does helping others really confer happiness or prosperity on the helper? I know of no evidence showing that altruists gain money from their altruism, but the evidence suggests that they often gain happiness. People who do volunteer work are happier and healthier than those who don't; but, as always, we have to contend with the problem of reverse correlation: Congenitally happy people are just plain nicer to begin with,[24] so their volunteer work may be a consequence of their happiness, not a cause. The happiness-as-cause hypothesis received direct support when the psychologist Alice Isen[25] went around Philadelphia leaving dimes in pay phones. The people who used those phones and found the dimes were then more likely to help a person who dropped a stack of papers (carefully timed to coincide with the phone caller's exit), compared with people who used phones that had empty coin-return slots. Isen has done more random acts of kindness than any other psychologist: She has distributed cookies, bags of candy, and packs of stationery; she has manipulated the outcome of video games (to let people win); and she has shown people happy pictures, always with the same finding: Happy people are kinder and more helpful than those in the control group.

What we need to find, however, is the reverse effect: that altruistic acts directly cause happiness and/or other long-term benefits. With its exhortation to "give blood; all you'll feel is good," is the American Red Cross telling the truth? The psychologist Jane Piliavin has studied blood donors in detail and found that, yes, giving blood does indeed make people feel good, and good about themselves. Piliavin[26] has reviewed the broader literature on all kinds of volunteer work and reached the conclusion that helping others does help the self, but in complex ways that depend on one's life stage. Research on "service learning," in which (mostly) high school students do volunteer work and engage in group reflection on what they are doing as part of a course, provides generally encouraging results: reduced delinquency and behavioral problems, increased civic participation, and increased commitment to positive social values. However, these programs do not appear to have much effect on the self-esteem or happiness of the adolescents involved. For adults, the story is a bit different. A longitudinal study[27] that tracked volunteering and well-being over many years in thousands of people was able to show a causal effect: When a person increased volunteer work, all measures of happiness and well-being increased (on average) afterwards, for as long as the volunteer work was a part of the person's life. The elderly benefit even more than do other adults, particularly when their volunteer work either involves direct person-to-person helping or is done through a religious organization. The benefits of volunteer work for the elderly are so large that they even show up in improved health and longer life. Stephanie Brown and her colleagues at the University of Michigan found striking evidence of such effects when they examined data from a large longitudinal study of older married couples.[28] Those who reported *giving* more help and support to spouses, friends, and relatives went on to live longer than those who gave less (even after controlling for factors such as health at the beginning of the study period), whereas the amount of help that people reported *receiving* showed no relationship to longevity. Brown's finding shows directly that, at least for older people, it really is more blessed to give than to receive.

This pattern of age-related change suggests that two of the big benefits of volunteer work are that it brings people together, and it helps them to construct a McAdams-style life story.[29] Adolescents are already immersed in a dense network of social relationships, and they are just barely beginning to construct their life stories, so they don't much need either of these benefits.

With age, however, one's story begins to take shape, and altruistic activities add depth and virtue to one's character. In old age, when social networks are thinned by the deaths of friends and family, the social benefits of volunteering are strongest (and indeed, it is the most socially isolated elderly who benefit the most from volunteering).[30] Furthermore, in old age, generativity, relationship, and spiritual strivings come to matter more, but achievement strivings seem out of place,[31] more appropriate for the middle chapters of a life story; therefore, an activity that lets one "give something back" fits right into the story and helps to craft a satisfying conclusion.

## THE FUTURE OF VIRTUE

Scientific research supports the virtue hypothesis, even when it is reduced to the claim that altruism is good for you. When it is evaluated in the way that Ben Franklin meant it, as a claim about virtue more broadly, it becomes so profoundly true that it raises the question of whether cultural conservatives are correct in their critique of modern life and its restricted, permissive morality. Should we in the West try to return to a more virtue-based morality?

I believe that we have indeed lost something important—a richly textured common ethos with widely shared virtues and values. Just watch movies from the 1930s and 1940s and you'll see people moving around in a dense web of moral fibers: Characters are concerned about their honor, their reputation, and the appearance of propriety. Children are frequently disciplined by adults other than their parents. The good guys always win, and crime never pays. It may sound stuffy and constraining to us now, but that's the point: Some constraint is good for us; absolute freedom is not. Durkheim, the sociologist who found that freedom from social ties is correlated with suicide[32] also gave us the word "anomie" (normlessness). Anomie is the condition of a society in which there are no clear rules, norms, or standards of value. In an anomic society, people can do as they please; but without any clear standards or respected social institutions to enforce those standards, it is harder for people to find things they want to do. Anomie breeds feelings of rootlessness and anxiety and leads to an increase in amoral and antisocial behavior. Modern sociological research strongly supports Durkheim: One of the best predictors of the health of an American neighborhood is the degree to which adults respond to the misdeeds of other people's children.[33] When community standards are enforced, there is constraint and cooperation. When everyone minds his own business and looks the other way, there is freedom and anomie.

My colleague at the University of Virginia, the sociologist James Hunter, carries Durkheim's ideas forward into the current debate about character education. In his provocative book *The Death of Character*,[34] Hunter traces out how America lost its older ideas about virtue and character. Before the Industrial Revolution, Americans honored the virtues of "producers"—hard work, self-restraint, sacrifice for the future, and sacrifice for the common good. But during the twentieth century, as people became wealthier and the producer society turned gradually into the mass consumption society, an alternative vision of the self arose—a vision centered on the idea of individual preferences and personal fulfillment. The intrinsically moral term "character" fell out of favor and was replaced by the amoral term "personality."

Hunter points to a second cause of character's death: inclusiveness. The first American colonists created enclaves of ethnic, religious, and moral homogeneity, but the history of America ever since has been one of increasing diversity. In response, educators have struggled to identify the ever-shrinking set of moral ideas everyone could agree upon. This shrinking reached its logical conclusion in the 1960s with the popular "values clarification" movement, which taught no morality at all. Values clarification taught children how to find their own values, and it urged teachers to refrain from imposing values on anyone. Although the goal of inclusiveness was laudable, it had unintended side effects: It cut children off from the soil of tradition, history, and religion that nourished older conceptions of virtue. You can grow vegetables hydroponically, but even then you have to add nutrients to the water. Asking children to grow virtues hydroponically, looking only within themselves for guidance, is like asking each one to invent a personal language—a pointless and isolating task if there is no community with whom to speak. (For a sensitive analysis from a more liberal perspective of the need for "cultural resources" for identity creation, see Anthony Appiah's *The Ethics of Identity*.)[35]

I believe Hunter's analysis is correct, but I am not yet convinced that we are worse off, overall, with our restricted modern morality. One thing that often distresses me in old movies and television programs, even up through the 1960s, is how limited were the lives of women and African Americans. We have paid a price for our inclusiveness, but we have bought ourselves a more humane society, with greater opportunity for racial minorities, women, gay people, the handicapped, and others—that is, for most people. And even if some people think the price was too steep, we can't go back, either to a pre-consumer society or to ethnically homogeneous enclaves. All we can do is search for ways that we might reduce our anomie without excluding large classes of people.

Being neither a sociologist nor an expert in education policy, I will not try to design a radical new approach to moral education. Instead, I will present one finding from my own research on diversity. The word "diversity" took on its current role in American discourse only after a 1978 Supreme Court ruling (U.C. Regents v. Bakke) that the use of racial preferences to achieve racial quotas at universities was unconstitutional, but that it was permissible to use racial preferences to increase diversity in the student body. Since then, diversity has been widely celebrated, on bumper stickers, in campus diversity days, and in advertisements. For many liberals, diversity has become an unquestioned good—like justice, freedom, and happiness, the more diversity, the better.

My research on morality, however, spurred me to question it. Given how easy it is to divide people into hostile groups based on trivial differences,[36] I wondered whether celebrating diversity might also encourage division, whereas celebrating commonality would help people form cohesive groups and communities. I quickly realized that there are two main kinds of diversity—demographic and moral. Demographic diversity is about sociodemographic categories such as race, ethnicity, sex, sexual orientation, age, and handicapped status. Calling for demographic diversity is in large measure calling for justice, for the inclusion of previously excluded groups. Moral diversity, on the other hand, is essentially what Durkheim described as anomie: a lack of consensus on moral norms and values. Once you make this distinction, you see that nobody can coherently even *want* moral diversity. If you are pro-choice on the issue of abortion, would you *prefer* that there be a wide variety of opinions and no dominant one? Or would you prefer that everyone agree with you and the laws of the land reflect that agreement? If you prefer diversity on an issue, the issue is not a moral issue for you; it is a matter of personal taste.

With my students Holly Hom and Evan Rosenberg, I conducted a study among several groups at the University of Virginia.[37] We found that there was strong support among students for increasing diversity for demographic categories (such as race, religion, and social class), even among students who described themselves as politically conservative. Moral diversity (opinions about controversial political questions), however, was much less appealing in most contexts, with the interesting exception of seminar classes. Students wanted to be exposed to moral diversity in class, but not in the people they live with and socialize with. Our conclusion from this study is that diversity is like cholesterol: There's a good kind and a bad kind, and perhaps we should not be trying to maximize both. Liberals are right to work for a society that is open to people of every demographic group, but conservatives might be right in believing that at the same time we should work much harder to create a common, shared identity. Although I am a political liberal, I believe that conservatives have a better understanding of moral development (although not of moral psychology in general—they are too committed to the myth of pure evil). Conservatives want schools to teach lessons that will create a positive and uniquely American identity, including a heavy dose of American history and civics, using English as the only national language. Liberals are justifiably wary of jingoism, nationalism, and the focus on books by "dead white males," but I think everyone who cares about education should remember that the American motto of *e pluribus, unum* (from many, one) has two parts. The celebration of *pluribus* should be balanced by policies that strengthen the *unum*.

Maybe it's too late. Maybe in the hostility of the current culture war, no one can find any value in the ideas of the other side. Or maybe we can turn for instruction to that great moral exemplar, Ben Franklin. Reflecting upon the way history is driven forward by people and parties fighting each other bitterly in pursuit of their self-interest, Franklin proposed creating a "United Party for Virtue." This party, composed of people who had cultivated virtue in themselves, would act only "with a view to the good of mankind." Perhaps that was naive even in Franklin's day, and it seems unlikely that these "good and wise men" would find it as easy to agree on a platform as Franklin supposed. Nonetheless, Franklin may be right that leadership on virtue can never come from the major political actors; it will have to come from a movement of people, such as the people of a town who come together and agree to create moral coherence across the many areas of children's lives. Such movements are happening now. The developmental psychologist William Damon[38] calls them "youth charter" movements, for they involve the cooperation of all parties to childrearing—parents, teachers, coaches, religious leaders, and the children themselves—who come to consensus on a "charter" describing the community's shared understandings, obligations, and values and committing all parties to expect and uphold the same high standards of behavior in all settings. Maybe youth charter communities can't rival the moral richness of ancient Athens, but they are doing something to reduce their own anomie while far exceeding Athens in justice.

# 9

## Divinity With or Without God

*We must not allow the ignoble to injure the noble, or the smaller to injure the greater. Those who nourish the smaller parts will become small men. Those who nourish the greater parts will become great men.*

— MENG TZU,[1] 3RD CENT. BCE

*God created the angels from intellect without sensuality, the beasts from sensuality without intellect, and humanity from both intellect and sensuality. So when a person's intellect overcomes his sensuality, he is better than the angels, but when his sensuality overcomes his intellect, he is worse than the beasts.*

— MUHAMMAD[2]

OUR LIFE IS THE CREATION of our minds, and we do much of that creating with metaphor. We see new things in terms of things we already understand: Life is a journey, an argument is a war, the mind is a rider on an elephant. With the wrong metaphor we are deluded; with no metaphor we are blind.

The metaphor that has most helped me to understand morality, religion, and the human quest for meaning is *Flatland*, a charming little book written in 1884 by the English novelist and mathematician Edwin Abbot.[3] Flatland is a two-dimensional world whose inhabitants are geometric figures. The protagonist is a square. One day, the square is visited by a sphere from a three-dimensional world called Spaceland. When a sphere visits Flatland, however, all that is visible to Flatlanders is the part of the sphere that lies in their plain—in other words, a circle. The square is astonished that the circle is able to grow or shrink at will (by rising or sinking into the plane of Flatland) and even to disappear and reappear in a different place (by leaving the plane, and then reentering it). The sphere tries to explain the concept of the third dimension to the two-dimensional square, but the square, though skilled at two-dimensional geometry, doesn't get it. He cannot understand what it means to have thickness in addition to height and breadth, nor can he understand that the circle came from up above him, where "up" does not mean from the north. The sphere presents analogies and geometrical demonstrations of how to move from one dimension to two, and then from two to three, but the square still finds the idea of moving "up" out of the plane of Flatland ridiculous.

In desperation, the sphere yanks the square up out of Flatland and into the third dimension so that the square can look down on his world and see it all at once. He can see the inside of all the houses and the guts (insides) of all the inhabitants. The square recalls the experience:

An unspeakable horror seized me. There was darkness; then a dizzy, sickening sensation of sight that was not like seeing; I saw space that was not space: I was myself, and not myself. When I could find voice, I shrieked aloud in agony, "Either this is madness or it is Hell." "It is neither," calmly replied the voice of the sphere, "it is Knowledge; it is Three Dimensions: open your eye once again and try to look steadily." I looked, and, behold, a new world!

The square is awestruck. He prostrates himself before the sphere and becomes the sphere's disciple. Upon his return to Flatland, he struggles to preach the "Gospel of Three Dimensions" to his fellow two-dimensional creatures—but in vain.

We are all, in some way, the square before his enlightenment. We have all encountered something we failed to understand, yet smugly believed we understood because we couldn't conceive of the dimension to which we were blind. Then one day something happens that makes no sense in our two-dimensional world, and we catch our first glimpse of another dimension.

In all human cultures, the social world has two clear dimensions: a horizontal dimension of closeness or liking, and a vertical one of hierarchy or status. People naturally and effortlessly make distinctions along the horizontal dimension between close versus distant kin, and between friends versus strangers. Many languages have one form of address for those who are close (*tu*, in French) and another for those who are distant (*vous*). We also have a great deal of innate mental structure that prepares us for hierarchical interactions. Even in hunter-gatherer cultures that are in many ways egalitarian, equality is only maintained by active suppression of ever-present tendencies toward hierarchy.[4] Many languages use the same verbal methods to mark hierarchy as they do to mark closeness (in French, *tu* for subordinates as well as friends, *vous* for superiors as well as strangers). Even in languages such as English that do not have different verb forms for different social relationships, people find a way to mark them anyway: We address people who are distant or superior by using their titles and last names (Mr. Smith, Judge Brown), and use first names for those who are intimate or subordinate.[5] Our minds automatically keep track of these two dimensions. Think how awkward it was the last time someone you barely knew but greatly revered invited you to call him by first name. Did the name stick in your throat? Conversely, when a salesperson addresses you by first name without having been invited to do so, do you feel slightly offended?

Now imagine yourself happily moving around your two-dimensional social world, a flat land where the X axis is closeness and the Y axis is hierarchy (see figure 9.1). Then one day, you see a person do something extraordinary, or you have an overwhelming experience of natural beauty, and you feel lifted "up." But it's not the "up" of hierarchy, it's some other kind of elevation. This chapter is about that vertical movement. My claim is that the human mind perceives a third dimension, a specifically moral dimension that I will call "divinity." (See the Z axis, coming up out of the plane of the page in figure 9.1). In choosing the label "divinity," I am not assuming that God exists and is there to be perceived. (I myself am a Jewish atheist.) Rather, my research on the moral emotions has led me to conclude that the

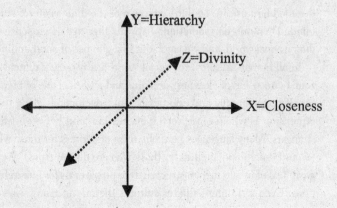

Fig. 9.1    The Three Dimensions of Social Space

human mind simply *does* perceive divinity and sacredness, whether or not God exists. In reaching this conclusion, I lost the smug contempt for religion that I felt in my twenties.

This chapter is about the ancient truth that devoutly religious people grasp, and that secular thinkers often do not: that by our actions and our thoughts, we move up and down on a vertical dimension. In the opening epigraph of this chapter, Meng Tzu called it a dimension of noble versus ignoble. Muhammad, like Christians and Jews before him, made it a dimension of divinity, with angels above and beasts below. An implication of this truth is that we are impoverished as human beings when we lose sight of this dimension and let our world collapse into two dimensions. But at the other extreme, the effort to create a three-dimensional society and impose it on all residents is the hallmark of religious fundamentalism. Fundamentalists, whether Christian, Jewish, Hindu, or Muslim, want to live in nations whose laws are in harmony with—or are taken from—a particular holy book. There are many reasons for democratic Western societies to oppose such fundamentalism, but I believe that the first step in such opposition must be an honest and respectful understanding of its moral motives. I hope that this chapter contributes to such understanding.

## ARE WE NOT ANIMALS?

I first found divinity in disgust. When I began to study morality, I read the moral codes of many cultures, and the first thing I learned is that most cultures are very concerned about food, sex, menstruation, and the handling of corpses. Because I had always thought morality was about how people treat each other, I dismissed all this stuff about "purity" and "pollution" (as the anthropologists call it) as extraneous to real morality. Why are women in many cultures forbidden to enter temples or touch religious artifacts while they are menstruating, or for a few weeks after giving birth?[6] It must be some sort of sexist effort to control women. Why is eating pork an abomination for Jews and Muslims? Must be a health-related effort to avoid trichinosis. But as I read further, I began to discern an underlying logic: the logic of disgust. According to the leading theory of disgust in the 1980s, by Paul Rozin,[7] disgust is largely about animals and the products of animal bodies (few plants or inorganic materials are disgusting), and disgusting things are contagious by touch. Disgust therefore seemed somehow related to the concerns about animals, body products (blood, excrement), washing, and touch that are so clear in the Old Testament, the Koran, Hindu scriptures, and many ethnographies of traditional societies. When I went to talk to Rozin about the possible role of disgust in morality and religion, I found that he had been thinking about the same question. With Professor Clark McCauley of Bryn Mawr College, we began to study disgust and the role it plays in social life.

Disgust has its evolutionary origins in helping people decide what to eat.[8] During the evolutionary transition in which our ancestors' brains expanded greatly, so did their production of tools and weapons, and so did their consumption of meat.[9] (Many scientists think these changes were all interrelated, along with the greater interdependence of male and female that I discussed in chapter 6). But when early humans went for meat, including scavenging the carcasses left by other predators, they exposed themselves to a galaxy of new microbes and parasites, most of which are contagious in a way that plant toxins are not: If a poisonous berry brushes up against your baked potato, it won't make the potato harmful or disgusting. Disgust was originally shaped by natural selection as a guardian of the mouth: It gave an advantage to individuals who went beyond the sensory properties of a potentially edible object (does it smell good?) and thought about where it came from and what it had touched. Animals that routinely eat or crawl on corpses, excrement, or garbage piles (rats, maggots, vultures, cockroaches) trigger disgust in us: We won't eat them, and anything they have touched becomes contaminated. We're also disgusted by most of the body products of other people, particularly excrement, mucus, and blood, which may transmit diseases among people. Disgust extinguishes desire (hunger) and motivates purifying behaviors such as washing or, if it's too late, vomiting.

But disgust doesn't guard just the mouth; its elicitors expanded during biological and cultural evolution so that now it guards the body more generally.[10] Disgust plays a role in sexuality analogous to its role in food selection by guiding people to the narrow class of culturally acceptable sexual partners and sexual acts. Once again, disgust turns off desire and motivates concerns about purification, separation, and cleansing. Disgust also gives us a queasy feeling when we see people with skin lesions, deformities, amputations, extreme obesity or thinness, and other violations of the culturally ideal outer envelope of the human body. It is the exterior that matters: Cancer in the lungs or a missing kidney is not disgusting; a tumor on the face or a missing finger is.

This expansion, from guardian of the mouth to guardian of the body, makes sense from a purely biological perspective: We humans have always lived in larger, denser groups than most other primates, and we lived on the ground, too, not in trees, so we were more exposed to the ravages of microbes and parasites that spread by physical contact. Disgust makes us careful about contact. But the most fascinating thing about disgust is that it is recruited to support so many of the norms, rituals, and beliefs that cultures use to define themselves.[11] For example, many cultures draw a sharp line between humans and animals, insisting that people are somehow above, better than, or more god-like than other animals. The human body is often thought of as a temple that houses divinity within: "Or do you not know that your body is a temple of the Holy Spirit within you, which you have from God, and that you are not your own? . . . [T]herefore glorify God in your body" (1 CORINTHIANS 6:19–20).

Yet a culture that says that humans are not animals, or that the body is a temple, faces a big problem: Our bodies do all the same things that animal bodies do, including eating, defecating, copulating, bleeding, and dying. The overwhelming evidence is that we *are* animals, and so a culture that rejects our animality must go to great lengths to hide the evidence. Biological processes must be carried out in the right way, and disgust is a guardian of that rightness. Imagine visiting a town where people wear no clothes, never bathe, have sex "doggie-style" in public, and eat raw meat by biting off pieces directly from the carcass. Okay, perhaps you'd pay to see such a freak show, but as with all freak shows, you would emerge degraded (literally: brought *down*). You would feel disgust at this "savage" behavior and know, viscerally, that there was something wrong with these people. Disgust is the guardian of the temple of the body. In this imaginary town, the guardians have been murdered, and the temples have gone to the dogs.

The idea that the third dimension—divinity—runs from animals below to god(s) above, with people in the middle, was perfectly captured by the seventeenth-century New England Puritan Cotton Mather, who observed a dog urinating at the same time he himself was urinating. Overwhelmed with disgust at the vileness of his own urination, Mather wrote this resolution in his diary: "Yet I will be a more noble creature; and at the very time when my natural necessities debase me into the condition of the beast, my spirit shall (I say at that very time!) rise and soar."[12]

If the human body is a temple that sometimes gets dirty, it makes sense that "cleanliness is next to Godliness."[13] If you don't perceive this third dimension, then it is not clear why God would care about the amount of dirt on your skin or in your home. But if you do live in a three-dimensional world, then disgust is like Jacob's ladder: It is rooted in the earth, in our biological necessities, but it leads or guides people toward heaven—or, at least, toward something felt to be, somehow, "up."

## THE ETHIC OF DIVINITY

After graduate school, I spent two years working with Richard Shweder, a psychological anthropologist at the University of Chicago who is the leading thinker in the field of cultural psychology. Shweder does much of his re-search in the Indian city of Bhubaneswar, in the state of Orissa, on the Bay of Bengal. Bhubaneswar is an ancient temple town—its old city grew up around the gigantic and ornate Lingaraj temple, built in the seventh century and still a major pilgrimage center for Hindus. Shweder's research on morality[14] in Bhubaneswar and elsewhere shows that when people think about morality, their moral concepts cluster into three groups, which he calls the ethic of autonomy, the ethic of community, and the ethic of divinity. When people think and act using the ethic of autonomy, their goal is to protect individuals from harm and grant them the maximum degree of autonomy, which they can use to pursue their own goals. When people use the ethic of community, their goal is to protect the integrity of groups, families, companies, or nations, and they value virtues such as obedience, loyalty, and wise leadership. When people use the ethic of divinity, their goal is to protect from degradation the divinity that exists in each person, and they value living in a pure and holy way, free from moral pollutants such as lust, greed, and hatred. Cultures vary in their relative reliance on these three ethics, which correspond, roughly, to the X, Y, and Z axes of figure 9.1. In my dissertation research[15] on moral judgment in Brazil and the United States, I found that educated Americans of high social class relied overwhelmingly on the ethic of autonomy in their moral discourse, whereas Brazilians, and people of lower social class in both countries, made much greater use of the ethics of community and divinity.

To learn more about the ethic of divinity, I went to Bhubaneswar for three months in 1993, to interview priests, monks, and other experts on Hindu worship and practice. To prepare, I read everything I could about Hinduism and the anthropology of purity and pollution, including *The Laws of Manu*,[16] a guidebook for Brahmin men (the priestly caste) written in the first or second century. Manu tells Brahmins how to live, eat, pray, and interact with other people while still attending to what Cotton Mather called their "natural necessities." In one passage, Manu lists the times when a priest should "not even think about" reciting the holy vedas (scriptures):

> while expelling urine or excrement, when food is still left on his mouth
> and hands, while eating at a ceremony for the dead, . . . when one has

eaten flesh or the food of a woman who has just given birth, . . . when jackals howl, . . . in a cremation ground, . . . while wearing a garment that he has worn in sexual union, while accepting anything at a cere- mony for the dead, when one has just eaten or has not digested (his food) or has vomited or belched, . . . when blood flows from one's limbs or when one has been wounded by a weapon.

This passage is extraordinary because it lists every category of disgust that Rozin, McCauley, and I had studied: food, body products, animals, sex, death, body envelope violations, and hygiene. Manu is saying that the presence in *mind* of the holy vedas is not compatible with contamination of the *body* from any source of disgust.[17] Divinity and disgust must be kept separate at all times.

When I arrived in Bhubaneswar, I quickly found that the ethic of divin- ity is not just ancient history. Even though Bhubaneswar is physically flat, it has a highly variable spiritual topography with peaks at each of its hun- dreds of temples. As a non-Hindu, I was allowed into the courtyards of temple compounds; and if I removed my shoes and any leather items (leather is polluting), I could usually enter the antechamber of the temple building. I could look into the inner sanctum where the god was housed, but had I crossed the threshold to join the Brahmin priest within, I would have polluted it and offended everyone. At the highest peak of divinity— the Lingaraj temple itself—I was not even allowed to enter the compound, although foreigners were invited to look in from an observation platform just outside the walls. It was not a matter of secrecy; it was a matter of con- tamination by people such as me who had not followed the proper proce- dures of bathing, diet, hygiene, and prayer for maintaining religious purity.

Hindu homes in Bhubaneswar have the same concentric structure as the temples: Leave your shoes at the door, socialize in the outer rooms, but never go into the kitchen or the room or area where offerings are made to deities. These two areas are maintained as zones of the highest purity. Even the human body has peaks and valleys, the head and the right hand being pure, the left hand and the feet being polluted. I had to take extraordinary care to keep my feet from touching anyone and to avoid handing something to another person with my left hand. As I moved around Bhubaneswar, I felt like a square in Spaceland as I tried to navigate a three-dimensional world with only the dimmest perception of its third dimension.

The interviews I conducted helped me to see a little better. My goal was to find out whether purity and pollution were really just about keeping biological "necessities" separate from divinity, or whether these practices had a deeper relationship to virtue and morality. I found a variety of opinions. Some of the less-educated village priests saw the rituals related to purity and pollution as basic rules of the game, things you simply must do because religious tradition demands it. But many of the people I interviewed took a broader view and saw purity and pollution practices as means to an end: spiritual and moral advancement, or moving up on the third dimension. For example, when I asked why it was important to guard one's purity, the headmaster of a Sanskrit school (a school that trains religious scholars) responded in this way:

> We ourselves can be gods or demons. It depends on karma. If a person behaves like a demon, for example he kills someone, then that person is truly a demon. A person who behaves in a divine manner, because a person has divinity in him, he is like a god. . . . We should know that we are gods. If we think like gods we become like gods, if we think like demons we become like demons. What is wrong with being like a demon? What is going on nowadays, it is demonic. Divine behavior means not cheating people, not killing people. Complete character. You have divinity, you are a god.

The headmaster, who of course had not read Shweder, gave a perfect statement of the ethic of divinity. Purity is not just about the body, it is about the soul. If you know that you have divinity in you, you will act accordingly: You will treat people well, and you will treat your body as a temple. In so doing, you will accumulate good karma, and you will come back in your next life at a higher level—literally higher on the vertical dimension of divinity. If you lose sight of your divinity, you will give in to your baser motives. In so doing, you will accumulate bad karma, and in your next incarnation you will return at a lower level as an animal or a demon. This linkage of virtue, purity, and divinity is not uniquely Indian; Ralph Waldo Emerson said exactly the same thing:

> He who does a good deed is instantly ennobled. He who does a mean deed is by the action itself contracted. He who puts off impurity thereby puts on purity. If a man is at heart just, then in so far is he God.[18]

## SACRED INTRUSIONS

When I returned to Flatland (the United States), I didn't have to think about purity and pollution anymore. I didn't have to think about the second dimension—hierarchy—very much, either. American university culture has only mild hierarchy (students often address professors by first name) compared with most Indian settings. So in some ways my life was reduced to one dimension—closeness, and my behavior was constrained only by the ethic of autonomy, which allowed me to do whatever I wanted, as long as I didn't hurt anyone else.

Yet, once I had learned to see in three dimensions, I saw glimmers of divinity scattered all about. I began to feel disgust for the American practice of marching around one's own house—even one's bedroom—wearing the same shoes that, minutes earlier, had walked through city streets. I adopted the Indian practice of removing my shoes at my door, and asking visitors to do likewise, which made my apartment feel more like a sanctuary, a clean and peaceful space separated more fully than before from the outside world. I noticed that it felt wrong to bring certain books into the bathroom. I noticed that people often spoke about morality using a language of "higher" and "lower." I became aware of my own subtle feelings upon witnessing people behaving in sleazy or "degraded" ways, feelings that were more than just disapproval; they were feelings of having been brought "down" in some way myself.

In my academic work, I discovered that the ethic of divinity had been central to public discourse in the United States until the time of the World War I, after which it began to fade (except in a few places, such as the

American South—which also maintained racial segregation practices based on notions of physical purity). For example, advice aimed at young people in the Victorian era routinely spoke of purity and pollution. In a widely reprinted book from 1897 titled *What a Young Man Ought to Know,*[19] Sylvanus Stall devoted an entire chapter to "personal purity" in which he noted that

> God has made no mistake in giving man a strong sexual nature, but any young man makes a fatal mistake if he allows the sexual to dominate, to degrade, and to destroy that which is highest and noblest in his nature.

To guard their purity, Stall advised young men to avoid eating pork, masturbating, and reading novels. By the 1936 edition, this entire chapter had been removed.

The vertical dimension of divinity was so obvious to people in the Victorian age that even scientists referred to it. In a chemistry textbook from 1867, after describing methods of synthesizing ethyl alcohol, the author felt compelled to warn his young readers that alcohol has the effect of "dulling the intellectual operations and moral instincts; seeming to pervert and destroy all that is pure and holy in man, while it robs him of his highest attribute—reason."[20] In his 1892 book promoting Darwin's theory of evolution, Joseph Le Conte, a professor of geology at the University of California at Berkeley, practically quoted Meng Tzu and Muhammad: "Man is possessed of two natures—a lower, in common with animals, and a higher, peculiar to himself. The whole meaning of sin is the humiliating bondage of the higher to the lower."[21]

But as science, technology, and the industrial age progressed, the Western world became "desacralized." At least that's the argument made by the great historian of religion Mircea Eliade. In *The Sacred and the Profane,*[22] Eliade shows that the perception of sacredness is a human universal. Regardless of their differences, all religions have places (temples, shrines, holy trees), times (holy days, sunrise, solstices), and activities (prayer, special dancing) that allow for contact or communication with something otherworldly and pure. To mark off sacredness, all other times, places, and activities are defined as profane (ordinary, not sacred). The borders between the sacred and the profane must be carefully guarded, and that's what rules of purity and pollution are all about. Eliade says that the modern West is the first culture in human history that has managed to strip time and space of all sacredness and to produce a fully practical, efficient, and profane world. This is the world that religious fundamentalists find unbearable and are sometimes willing to use force to fight against.

Eliade's most compelling point, for me, is that sacredness is so irrepressible that it intrudes repeatedly into the modern profane world in the form of "crypto-religious" behavior. Eliade noted that even a person committed to a profane existence has

> privileged places, qualitatively different from all others—a man's birthplace, or the scenes of his first love, or certain places in the first foreign city he visited in his youth. Even for the most frankly nonreligious man, all these places still retain an exceptional, a unique quality; they are the "holy places" of his private universe, as if it were in such spots that he had received the revelation of a reality other than that in which he participates through his ordinary daily life.

When I read this, I gasped. Eliade had perfectly pegged my feeble spirituality, limited as it is to places, books, people, and events that have given me moments of uplift and enlightenment. Even atheists have intimations of sacredness, particularly when in love or in nature. We just don't infer that God caused those feelings.

## ELEVATION AND AGAPE

My time in India did not make me religious, but it did lead to an intellectual awakening. Shortly after moving to the University of Virginia in 1995, I was writing yet another article about how social disgust is triggered when we see people moving "down" on the vertical dimension of divinity. Suddenly it occurred to me that I had never really thought about the emotional reaction to seeing people move "up." I had referred in passing to the feeling of being "uplifted," but had never even wondered whether "uplift" is a real, honest-to-goodness emotion. I began to interrogate friends, family, and students: "When you see someone do a really good deed, do you feel something? What exactly? Where in your body do you feel it? Does it make you want to do anything?" I found that most people had the same feelings I did, and the same difficulty articulating exactly what they were. People talked about an open, warm, or glowing feeling. Some specifically mentioned the heart; others claimed they could not say where in their bodies they felt it, yet even as they were denying a specific location, their hands sometimes made a circular motion in front of the chest, fingers pointing inward as if to indicate something moving in the heart. Some people mentioned feelings of chills, or of choking up. Most said this feeling made them want to perform good deeds or become better in some way. Whatever this feeling was, it was beginning to look like an emotion worthy of study. Yet there was no research of any kind on this emotion in the psychological literature, which was focused at the time on the six "basic" emotions[23] known to have distinctive facial expressions: joy, sadness, fear, anger, disgust, and surprise.

If I believed in God, I would believe that he sent me to the University of Virginia for a reason. At UVA, a great deal of crypto-religious activity centers around Thomas Jefferson, our founder, whose home sits like a temple on a small mountaintop (Monticello) a few miles away. Jefferson wrote the holiest text of American history—the Declaration of Independence. He also wrote thousands of letters, many of which reveal his views on psychology, education, and religion. After arriving at UVA, having an Eliade-style crypto-religious experience at Monticello, and committing myself to the cult of Jefferson, I read a collection of his letters. There I found a full and perfect description of the emotion I had just begun thinking about.

In 1771, Jefferson's relative Robert Skipwith asked him for advice on what books to buy for the personal library he hoped to build. Jefferson, who loved giving advice almost as much as he loved books, happily obliged. Jefferson sent along a catalogue of serious works of history and philosophy, but he also recommended the purchase of fiction. In his day (as in Sylvanus Stall's), plays and novels were not regarded as worthy of a dignified man's time, but Jefferson justified his unorthodox advice by pointing out that great writing can trigger beneficial emotions:

When any ... act of charity or of gratitude, for instance, is presented
either to our sight or imagination, we are deeply impressed with its
beauty and feel a strong desire in ourselves of doing charitable and grateful acts also. On the contrary, when we see or read of any atrocious deed,
we are disgusted with its deformity, and conceive an abhorrence of vice.
Now every emotion of this kind is an exercise of our virtuous dispositions, and dispositions of the mind, like limbs of the body, acquire
strength by exercise.[24]

Jefferson went on to say that the physical feelings and motivational effects caused by great literature are as powerful as those caused by real
events. He considered the example of a contemporary French play, asking
whether the fidelity and generosity of its hero does not

dilate [the reader's] breast and elevate his sentiments as much as any
similar incident which real history can furnish? Does [the reader] not in
fact feel himself a better man while reading them, and privately covenant
to copy the fair example?

This extraordinary statement is more than just a poetic description of the
joys of reading. It is also a precise scientific definition of an emotion. In
emotion research, we generally study emotions by specifying their components, and Jefferson gives us most of the major components: an eliciting or
triggering condition (displays of charity, gratitude, or other virtues); physical
changes in the body ("dilation" in the chest); a motivation (a desire of "doing
charitable and grateful acts also"); and a characteristic feeling beyond bodily
sensations (elevated sentiments). Jefferson had described exactly the emotion I had just "discovered." He even said that it was the opposite of disgust.
As an act of crypto-religious glorification, I considered calling this emotion
"Jefferson's emotion," but thought better of it, and chose the word "elevation," which Jefferson himself had used to capture the sense of rising on a
vertical dimension, away from disgust.

For the past seven years I have been studying elevation in the lab. My
students and I have used a variety of means to induce elevation and have

found that video clips from documentaries about heroes and altruists, and selections from the Oprah Winfrey show, work well. In most of our studies, we show people in one group an elevating video, while people in the control condition see a video designed to amuse them, such as a Jerry Seinfeld monologue. We know (from Alice Isen's coins and cookies studies)[25] that feeling happy brings a variety of positive effects, so in our research we always try to show that elevation is not just a form of happiness. In our most comprehensive study,[26] Sara Algoe and I showed videos to research subjects in the lab and had them fill out a recording sheet about what they felt and what they wanted to do. Sara then gave them a stack of blank recording sheets and told them to keep an eye out, for the next three weeks, for instances of someone doing something good for someone else (in the elevation condition) or for times when they saw someone else tell a joke (in the amusement/control condition). We also added a third condition to study nonmoral admiration: People in this condition watched a video about the superhuman abilities of the basketball star Michael Jordan, and were then asked to record times when they witnessed someone doing something unusually skillful.

Both parts of Sara's study show that Jefferson got it exactly right. People really do respond emotionally to acts of moral beauty, and these emotional reactions involve warm or pleasant feelings in the chest and conscious desires to help others or become a better person oneself. A new discovery in Sara's study is that moral elevation appears to be different from admiration for nonmoral excellence. Subjects in the admiration condition were more likely to report feeling chills or tingles on their skin, and to report feeling energized or "psyched up." Witnessing extraordinarily skillful actions gives people the drive and energy to try to copy those actions.[27] Elevation, in contrast, is a calmer feeling, not associated with signs of physiological arousal. This distinction might help explain a puzzle about elevation. Although people say, in all our studies, that they *want* to do good deeds, in two studies where we gave them the opportunity to sign up for volunteer work or to help an experimenter pick up a stack of papers she had dropped, we did not find that elevation made people behave much differently.

What's going on here? How could an emotion that makes people rise on the dimension of divinity not make them behave more altruistically? It's too soon to know for sure, but a recent finding suggests that love could be the answer. Three undergraduate honors students have worked with me on the physiology of elevation—Chris Oveis, Gary Sherman, and Jen Silvers. We've all been intrigued by the frequency with which people who are feeling elevation point to the heart. We believe they're not just speaking metaphorically. Chris and Gary have found hints that the vagus nerve might be activated during elevation. The vagus nerve is the main nerve of the parasympathetic nervous system, which calms people down, and undoes the arousal caused by the sympathetic (fight-or-flight) system. The vagus nerve is the main nerve that controls heart rate, and it has a variety of other effects on the heart and lungs, so if people feel something in the chest, the vagus nerve is the main suspect, and it has already been implicated in research on feelings of gratitude and "appreciation."[28] But it's difficult to measure the activity of the vagus nerve directly, and so far Chris and Gary have found only hints, not conclusive proof.

Nerves have accomplices, however; they sometimes work with hormones to produce long-lasting effects, and the vagus nerve works with the hormone oxytocin to create feelings of calmness, love, and desire for contact that encourage bonding and attachment.[29] Jen Silvers was interested in the possible role of oxytocin in elevation, but because we did not have the resources to draw blood from subjects before and after watching an elevating video (which we'd have to do to detect a change in oxytocin levels), I told Jen to scour the research literature to find an indirect measure—something oxytocin does to people that we could measure without a hypodermic needle. Jen found one: lactation. One of oxytocin's many jobs in regulating the attachment of mothers and children is to trigger the release of milk in mothers who breast-feed.

In one of the boldest undergraduate honors theses ever done in the UVA psychology department, Jen brought forty-five lactating women into our lab (one at a time), with their babies, and asked them to insert nursing pads into their bras. Half the women then watched an elevating clip from an Oprah Winfrey show (about a musician who, after expressing his gratitude to the music teacher who had saved him from a life of gang violence, finds out that Oprah has brought in some of *his own* students to express their gratitude to him). The other mothers saw a video clip featuring several comedians. The women watched the videos in a private screening room, and a video camera (not hidden) recorded their behavior. When the videos were over, the mothers were left alone with their children for five minutes. At the end of the study, Jen weighed the nursing pads to measure milk release, and later coded the videos for whether the mothers nursed their babies or played warmly with them. The effect was one of the biggest I have ever found in any study: Nearly half of the mothers in the elevation condition either leaked milk or nursed their babies; only a few of the mothers in the comedy condition leaked or nursed. Furthermore, the elevated mothers showed more warmth in the way they touched and cuddled their babies. All of this suggests that oxytocin might be released during moments of elevation. And if this is true, then perhaps it was naive of me to expect that elevation would actually cause people to help strangers (even though they often say they want to do so). Oxytocin causes bonding, not action. Elevation may fill people with feelings of love, trust,[30] and openness, making them more receptive to new relationships; yet, given their feelings of relaxation and passivity, they might be less likely to engage in active altruism toward strangers.

The relationship of elevation to love and trust was beautifully expressed in a letter I once received from a man in Massachusetts, David Whitford, who had read about my work on elevation. Whitford's Unitarian church had asked each of its members to write a spiritual autobiography—an account of how each had become the spiritual person he or she is now. In one section of his autobiography, Whitford puzzled over why he was so often moved to tears during church services. He noticed that he shed two kinds of tears in church. The first he called "tears of compassion," such as the time he cried during a sermon on Mothers' Day on the subject of children who were abandoned or neglected. These cases felt to him like "being pricked in the soul," after which "love pours out" for those who are suffering. But he called the second kind "tears of celebration"; he could just as well have called them tears of elevation:

There's another kind of tear. This one's less about giving love and more about the joy of receiving love, or maybe just detecting love (whether it's directed at me or at someone else). It's the kind of tear that flows in response to expressions of courage, or compassion, or kindness by others. A few weeks after Mother's Day, we met here in the sanctuary after the service and considered whether to become a Welcoming Congregation [a congregation that welcomes gay people]. When John stood in support of the resolution, and spoke of how, as far as he knew, he was the first gay man to come out at First Parish, in the early 1970s, I cried for his courage. Later, when all hands went up and the resolution passed unanimously, I cried for the love expressed by our congregation in that act. That was a tear of celebration, a tear of receptiveness to what is good in the world, a tear that says it's okay, relax, let down your guard, there are good people in the world, there is good in people, love is real, it's in our nature. That kind of tear is also like being pricked, only now the love pours in.[31]

Growing up Jewish in a devoutly Christian country, I was frequently puzzled by references to Christ's love and love through Christ. Now that I understand elevation and the third dimension, I think I'm beginning to get it. For many people, one of the pleasures of going to church is the experience of collective elevation. People step out of their everyday profane existence, which offers only occasional opportunities for movement on the third dimension, and come together with a community of like-hearted people who are also hoping to feel a "lift" from stories about Christ, virtuous people in the Bible, saints, or exemplary members of their own community. When this happens, people find themselves overflowing with love, but it is not exactly the love that grows out of attachment relationships.[32] That love has a specific object, and it turns to pain when the object is gone. This love has no specific object; it is *agape*. It feels like a love of all humankind, and because humans find it hard to believe that something comes from nothing, it seems natural to attribute the love to Christ, or to the Holy Spirit moving within one's own heart. Such experiences give direct and subjectively compelling evidence that God resides within each person. And once a person knows this "truth," the ethic of divinity becomes self-evident. Some ways of living are compatible with divinity—they bring out the higher, nobler self; others do not. The split between the Christian left and the Christian right could be, in part, that some people see tolerance and acceptance as part of their nobler selves; others feel that they can best honor God by working to change society and its laws to conform to the ethic of divinity, even if that means imposing religious laws on people of other faiths.

## AWE AND TRANSCENDENCE

Virtue is not the only cause of movement on the third dimension. The vastness and beauty of nature similarly stirs the soul. Immanuel Kant explicitly linked morality and nature when he declared that the two causes of genuine awe are "the starry sky above and the moral law within."[33] Darwin felt spiritually uplifted while exploring South America:

> In my journal I wrote that whilst standing in midst of the grandeur of a Brazilian forest, "it is not possible to give an adequate idea of the higher feelings of wonder, admiration, and devotion which fill and elevate the mind." I well remember my conviction that there is more in man than the breath of his body.[34]

The New England transcendentalist movement was based directly on the idea that God is to be found in each person and in nature, so spending time alone in the woods is a way of knowing and worshiping God. Ralph Waldo Emerson, a founder of the movement, wrote:

> Standing on the bare ground—my head bathed by the blithe air and uplifted into infinite space—all mean egotism vanishes. I become a transparent eyeball; I am nothing; I see all; the currents of the Universal Being circulate through me; I am part or parcel of God. The name of the nearest friend sounds then foreign and accidental; to be brothers, to be acquaintances, master or servant, is then a trifle and a disturbance. I am the lover of uncontained and immortal beauty.[35]

Something about the vastness and beauty of nature makes the self feel small and insignificant, and anything that shrinks the self creates an opportunity for spiritual experience. In chapter 1, I wrote about the divided self—the many ways in which people feel as though they have multiple selves or intelligences that sometimes conflict. This division is often explained by positing a soul—a higher, noble, spiritual self, which is tied down to a body—a lower, base, carnal self. The soul escapes the body only at death; but before then, spiritual practices, great sermons, and awe at nature can give the soul a taste of the freedom to come.

There are many other ways of getting such a foretaste. People often refer to viewing great art, hearing a symphony, or listening to an inspiring speaker as (crypto) religious experiences. And some things give more than a taste: They give a full-blown, though temporary, escape. When the hallucinogenic drugs LSD and psilocybin became widely known in the West, medical researchers called these drugs "psychoto-mimetic" because they mimicked some of the symptoms of psychotic disorders such as schizophrenia. But those who tried the drugs generally rejected that label and made up terms such as "psychedelic" (manifesting the mind) and "entheogen" (generating God from within). The Aztec word for the psilocybin mushroom was *teonanacatl*, which means literally "god's flesh"; when it was eaten in religious ceremonies, it gave many the experience of a direct encounter with God.[36]

Drugs that create an altered mental state have an obvious usefulness in marking off sacred experiences from profane, and therefore many drugs, including alcohol and marijuana, play a role in religious rites in some cultures. But there is something special about the phenethylamines—the drug class that includes LSD and psilocybin. Drugs in this class, whether naturally occurring (as in psilocybin, mescaline, or yage) or synthesized by a chemist (LSD, ecstasy, DMT) are unmatched in their ability to induce massive alterations of perception and emotion that sometimes feel, even to secular users, like contact with divinity, and that cause people to feel afterwards that they've been transformed.[37] The effects of these drugs depend greatly on what Timothy Leary and the other early psychedelic explorers called "set and setting," referring to the user's mental set, and to the setting in which the drugs are taken. When people bring a reverential mindset and take the drugs in a safe and supportive setting, as is done in the initiation rites of some traditional cultures,[38] these drugs can be catalysts for spiritual and personal growth.

In the most direct test of this catalyst hypothesis, Walter Pahnke,[39] a physician working on a dissertation in theology, brought twenty graduate students in theology into a room below the chapel at Boston University on Good Friday 1962. He gave ten of the students 30 milligrams of psilocybin; the other ten were given identical-looking pills containing vitamin B5 (nicotinic acid), which creates feelings of tingles and flushing on the skin. The vitamin B5 is what's known as an active placebo: It creates real bodily feelings, so if the beneficial effects of psilocybin were just placebo effects, the control group would have good reason to show them. Over the next few hours, the whole group listened (via speakers) to the Good Friday service going on in the chapel upstairs. Nobody, not even Pahnke, knew who had taken which pill. But two hours after the pills were taken, there could be no doubt. Those who had taken the placebo were the first to feel something happening, and they assumed they had gotten the psilocybin. But nothing else happened. Half an hour later, the other students began an experience that many later described as one of the most important in their lives. Pahnke interviewed them after the drug wore off, and again a week later, and again six months later. He found that most of the people in the psilocybin group reported most of the nine features of mystical experience he had set out to measure. The strongest and most consistent effects included feelings of unity with the universe, transcendence of time and space, joy, a difficulty putting the experience into words, and a feeling of having been changed for the better. Many reported seeing beautiful colors and patterns and having profound feelings of ecstasy, fear, and awe.

Awe is *the* emotion of self-transcendence. My friend Dacher Keltner, an expert on emotion at the University of California at Berkeley, proposed to me a few years ago that we review the literature on awe and try to make sense of it ourselves. We found[40] that scientific psychology had almost nothing to say about awe. It can't be studied in other animals or created easily in the lab, so it doesn't lend itself to experimental research. But philosophers, sociologists, and theologians had a great deal to say about it. As we traced the word "awe" back in history, we discovered that it has always had a link to fear and submission in the presence of something much greater than the self. It's only in very modern times—in our de-sacralized world, perhaps—that awe has been reduced to surprise plus approval, and the word "awesome," much used by American teenagers, has come to mean little more than "double-plus good" (to use George Orwell's term from *1984*). Keltner and I concluded that the emotion of awe happens when two conditions are met: a person perceives something vast (usually physically vast, but sometimes conceptually vast, such as a grand theory; or socially vast, such as great fame or power); and the vast thing cannot be accommodated by the person's existing mental structures. Something enormous can't be processed, and when people are stumped, stopped in their cognitive tracks while in the presence of something vast, they feel small, powerless, passive, and receptive. They often (though not always) feel fear, admiration, elevation, or a sense of beauty as well. By stopping people and making them receptive, awe creates an opening for change, and this is why awe plays a role in most stories of religious conversion.

We found a prototype of awe—a perfect but extreme case—in the dramatic climax of the *Bhagavad Gita*. The *Gita* is an episode within the much longer story of the *Mahabharata,* an epic work about a war between two branches of an Indian royal family. As the hero of the story, Arjuna, is about to lead his troops into battle, he loses his nerve and refuses to fight. He does not want to lead his kinsmen into slaughter against his kinsmen. The *Gita* is the story of how Krishna (a form of the god Vishnu) persuades Arjuna that he must lead his troops into battle. In the middle of the battlefield, with troops arrayed on both sides, Krishna gives a detailed and abstract theological lecture on the topic of dharma—the moral law of the universe. Arjuna's dharma requires that he fight and win this war. Not surprisingly (given the weakness of reason when it comes to motivating action), Arjuna is unmoved. Arjuna asks Krishna to show him this universe of which he speaks. Krishna grants Arjuna's request and gives him a cosmic eye that allows him to see God and the universe as they really are. Arjuna then has an experience that sounds to modern readers like an LSD trip. He sees suns, gods, and infinite time. He is filled with amazement. His hair stands on end. He is disoriented and confused, unable to comprehend the wonders he is seeing. I don't know whether Edwin Abbot read the *Bhagavad Gita*, but the square's experience in Spaceland is exactly like Arjuna's. Arjuna is clearly in a state of awe when he says, "Things never before seen have I seen, and ecstatic is my joy; yet fear-and-trembling perturb my mind."[41] When the cosmic eye is removed and Arjuna comes "down" from his trip, he does just what the square did: He prostrates himself before the God who enlightened him, and he begs to serve. Krishna commands Arjuna to be loyal to him, and to cut off all other attachments. Arjuna gladly obeys, and, from then on, he honors Krishna's commands.

Arjuna's experience is extreme—the stuff of scripture; yet many people have had a spiritually transformative experience that included many of the same elements. In what is still the greatest work on the psychology of religion, William James analyzed the "varieties of religious experience,"[42] including rapid and gradual religious conversions and experiences with drugs and nature. James found such extraordinary similarity in the reports of these experiences that he thought they revealed deep psychological truths. One of the deepest truths, James said, was that we experience life as a divided self, torn by conflicting desires. Religious experiences are real and common, whether or not God exists, and these experiences often make people feel whole and at peace. In the rapid type of conversion experience (such as those of Arjuna and the square), the old self, full of petty concerns, doubts, and grasping attachments, is washed away in an instant, usually an instant of profound awe. People feel reborn and often remember the exact time and place of this rebirth, the moment they surrendered their will to a higher power and were granted direct experience of deeper truth. After such rebirth, fear and worry are greatly diminished and the world seems clean, new, and bright. The self is changed in ways that any priest, rabbi, or psychotherapist would call miraculous. James described these changes:

> The man who lives in his religious centre of personal energy, and is actuated by spiritual enthusiasms, differs from his previous carnal self in perfectly definite ways. The new ardor which burns in his breast consumes in its glow the lower "noes" which formerly beset him, and keeps him immune against infection from the entire groveling portion of his nature. Magnanimities once impossible are now easy; paltry conventionalities and mean incentives once tyrannical hold no sway. The stone wall inside of him has fallen, the hardness in his heart has broken down. The rest of us can, I think, imagine this by recalling our state of feeling in those temporary "melting moods" into which either the trials of real life, or the theatre, or a novel sometimes throw us. Especially if we weep! For it is then as if

our tears broke through an inveterate inner dam, and let all sorts of ancient peccancies and moral stagnancies drain away, leaving us now washed and soft of heart and open to every nobler leading.[43]

James's "melting moods" are strikingly similar to the feelings of elevation described by Jefferson and by David Whitford.

Atheists may protest that they, too, can have many of the same experiences without God. The psychologist who took such secular experiences seriously was Abraham Maslow, Harry Harlow's first graduate student and a founder of humanistic psychology. Maslow collected reports of what he called "peak experiences"—those extraordinary self-transcendent moments that feel qualitatively different from ordinary life. In a small gem of a book, *Religions, Values, and Peak Experiences,*[44] Maslow listed twenty-five common features of peak experiences, nearly all of which can be found somewhere in William James. Here are some: The universe is perceived as a unified whole where everything is accepted and nothing is judged or ranked; egocentrism and goal-striving disappear as a person feels merged with the universe (and often with God); perceptions of time and space are altered; and the person is flooded with feelings of wonder, awe, joy, love, and gratitude.

Maslow's goal was to demonstrate that spiritual life has a naturalistic meaning, that peak experiences are a basic fact about the human mind. In all eras and all cultures, many people have had these experiences, and Maslow suggested that all religions are based on the insights of somebody's peak experience. Peak experiences make people nobler, just as James had said, and religions were created as methods of promoting peak experiences and then maximizing their ennobling powers. Religions sometimes lose touch with their origins, however; they are sometimes taken over by people who have not had peak experiences—the bureaucrats and company men who want to routinize procedures and guard orthodoxy for orthodoxy's sake. This, Maslow said, is why many young people became disenchanted with organized religion in the mid-twentieth-century, searching instead for peak experiences in psychedelic drugs, Eastern religions, and new forms of Christian worship.

Maslow's analysis probably does not shock you. It makes sense as a secular psychological explanation of religion. But what is most surprising in *Religions, Values, and Peak Experiences* is Maslow's attack on science for becoming as sterile as organized religion. The historians of science Lorraine Daston and Katherine Park[45] later documented this change. They showed that scientists and philosophers had traditionally held an attitude of wonder toward the natural world and the objects of their inquiry. But in the late sixteenth century, European scientists began to look down on wonder; they began to see it as the mark of a childish mind, whereas the mature scientist went about coolly cataloging the laws of the world. Scientists may tell us in their memoirs about their private sense of wonder, but the everyday world of the scientist is one that rigidly separates facts from values and emotions. Maslow echoed Eliade in claiming that science has helped to de-sacralize the world, that it is devoted to documenting only what *is*, rather than what is *good* or what is *beautiful*. One might object that there is an academic division of labor; the good and the beautiful are the province of the humanities, not of the sciences. Maslow charged, however, that the humanities had abdicated their responsibility with their retreat to relativism, their skepticism about the possibility of truth, and their preference for novelty and iconoclasm over beauty. He founded humanistic psychology in part to feed the widespread hunger for knowledge about values and to investigate the sort of truth people glimpse in peak experiences. Maslow did not believe religions were literally true (as actual accounts of God and creation), but he thought they were based on the most important truths of life, and he wanted to unite those truths with the truths of science. His goal was nothing less than the reformation of education and, therefore, of society: "Education must be seen as at least partially an effort to produce the good human being, to foster the good life and the good society."[46]

## THE SATANIC SELF

The self is one of the great paradoxes of human evolution. Like the fire stolen by Prometheus, it made us powerful but exacted a cost. In *The Curse of the Self*,[47] the social psychologist Mark Leary points out that many other animals can think, but none, so far as we know, spend much time thinking about themselves. Only a few other primates (and perhaps dolphins) can even learn that the image in a mirror belongs to them.[48] Only a creature with language ability has the mental apparatus to focus attention on the self, to think about the self's invisible attributes and long term goals, to create a narrative about that self, and then to react emotionally to thoughts about that narrative. Leary suggests that this ability to create a self gave our ancestors many useful skills, such as long-term planning, conscious decision making and self-control, and the ability to see other people's perspectives. Because these skills are all important for enabling human beings to work closely together on large projects, the development of the self may have been crucial to the development of human ultrasociality. But by giving each one of us an inner world, a world full of simulations, social comparisons, and reputational concerns, the self also gave each one of us a personal tormenter. We all now live amid a whirlpool of inner chatter, much of which is negative (threats loom larger than opportunities), and most of which is useless. It is important to note that the self is not exactly the rider—much of the self is unconscious and automatic—but because the self emerges from conscious verbal thinking and storytelling, it can be constructed only by the rider.

Leary's analysis shows why the self is a problem for all major religions: The self is the main obstacle to spiritual advancement, in three ways. First, the constant stream of trivial concerns and egocentric thoughts keeps people locked in the material and profane world, unable to perceive sacredness and divinity. This is why Eastern religions rely heavily on meditation, an effective means of quieting the chatter of the self. Second, spiritual transformation is essentially the transformation of the self, weakening it, pruning it back—in some sense, killing it—and often the self objects. Give up my possessions and the prestige they bring? No way! Love my enemies, after what they did to me? Forget about it. And third, following a spiritual path is invariably hard work, requiring years of meditation, prayer, self-control, and sometimes self-denial. The self does not like to be denied, and it is adept at finding reasons to bend the rules or cheat. Many religions teach that egoistic attachments to pleasure and reputation are constant temptations to leave the path of virtue. In a sense, the self is Satan, or, at least, Satan's portal.

For all these reasons, the self is a problem for the ethic of divinity. The big greedy self is like a brick holding down the soul. Only by seeing the self in this way, I believe, can one understand and even respect the moral motivations of those who want to make their society conform more closely to the particular religion they follow.

## FLATLAND AND THE CULTURE WAR

Humor helps people cope with adversity, and after George W. Bush received a majority of the votes in the U.S. presidential election of 2004, 49 percent of Americans had a lot of coping to do. Many people in the "blue states" (those where a majority voted for John Kerry, shown on all electoral maps in blue) could not understand why people in the "red states" supported Bush and his policies. Liberals posted maps of the United States on the Internet that showed the blue states (all in the Northeast, the upper Midwest, and along the West coast) labeled "United States of America"; the red states (almost the whole interior and south of the nation) were labeled "Jesusland." Conservatives countered with their own map in which the blue states were labeled "New France," but I think a more accurate parody, from the right's point of view, might have been to call the blue states "Selfland."

I am not suggesting that people who voted for John Kerry are any more selfish than those who voted for George Bush—indeed, the taxation and social policies of the two candidates suggest just the opposite. But I am trying to understand the mutual incomprehension of the two sides in the culture war, and I believe that Shweder's three ethics—particularly the ethic of divinity—are the key to it.

Which of the following quotations inspires you more: (1) "Self-esteem is the basis of any democracy"; (2) "It's not all about you." The first is attributed to Gloria Steinem,[49] a founder of the feminist movement in the 1970s. It claims that sexism, racism, and oppression make particular groups of people feel unworthy and therefore undermine their participation in democracy. This quote also reflects the core idea of the ethic of autonomy: Individuals are what really matter in life, so the ideal society protects all individuals from harm and respects their autonomy and freedom of choice. The ethic of autonomy is well suited to helping people with different backgrounds and values get along with each other because it allows each person to pursue the life she chooses, as long as those choices don't interfere with the rights of others.

The second quote is the opening line of the world's biggest-selling book in 2003 and 2004, *The Purpose Driven Life* by Rick Warren,[50] a guide for finding purpose and meaning through faith in Jesus Christ and the revelation of the Bible. From Warren's perspective, the self is the cause of our problems and therefore efforts to raise children's self-esteem directly with awards, praise, and exercises to make them feel "special" are positively evil. The core idea of the ethic of divinity is that each person has divinity inside, so the ideal society helps people live in a way consistent with that divinity. What an individual desires is not particularly important—many desires come from the carnal self. Schools, families, and the media should all work together to help children overcome their sense of self and entitlement and live instead in the way Christ intended.

Many of the key battles in the American culture war are essentially about whether some aspect of life should be structured by the ethic of autonomy or by the ethic of divinity.[51] (The ethic of community, which stresses the importance of the group over that of the individual, tends to be allied with the ethic of divinity). Should there be prayer in schools? Should the Ten Commandments be posted in schools and courthouses? Should the phrase "under God" be struck from the American pledge of allegiance? Liberals usually want to keep religion out of public life so that people cannot be forced to participate against their will, but religious conservatives want schools and courthouses re-sacralized. They want their children to live in a (particular) three-dimensional world, and if the schools won't provide it, they sometimes turn to home-schooling instead.

Should people be allowed to use birth control, abortion, reproductive technologies, and assisted suicide as they please? It depends on whether your goal is to empower people to manage some of the most important choices of their lives, or whether you think all such decisions must be made by God. If the book title *Our Bodies, Ourselves* sounds like a noble act of defiance to you, you will support people's rights to choose their own sexual activities and to modify their bodies as they please. But if you believe that "God prescribed every single detail of your body,"[52] as Warren writes in *The*

*Purpose Driven Life,* you will probably be offended by sexual diversity and by body modifications such as piercings and plastic surgery. My students and I have interviewed political liberals and conservatives about sexual morality,[53] and about body modifications,[54] and in both studies we found that liberals were much more permissive and relied overwhelmingly on the ethic of autonomy; conservatives, much more critical, used all three ethics in their discourse. For example, one conservative man justified his condemnation of a story about an unusual form of masturbation:

> It's a sin because it distances ourselves from God. It's a pleasure that God did not design for us to enjoy because sexual pleasures, through, you know, a married heterosexual couple, were designed by God in order to reproduce.[55]

On issue after issue, liberals want to maximize autonomy by removing limits, barriers, and restrictions. The religious right, on the other hand, wants to structure personal, social, and political relationships in three dimensions and so create a landscape of purity and pollution where restrictions maintain the separation of the sacred and the profane. For the religious right, hell on earth is a flat land of unlimited freedom where selves roam around with no higher purpose than expressing and developing themselves.

As a liberal, I value tolerance and openness to new ideas. I have done my best, in this chapter, to be tolerant toward those whose politics I oppose and to find merit in religious ideas I do not hold. But although I have begun to see the richness that divinity adds to human experience, I do not entirely lament the "flattening" of life in the West over the last few hundred years. An unfortunate tendency of three-dimensional societies is that they often include one or more groups that get pushed down on the third dimension and then treated badly, or worse. Look at the conditions of "untouchables" in India until recently, or at the plight of Jews in medieval Europe and in purity-obsessed Nazi Germany, or at the humiliation of African Americans in the segregated South. The American religious right now seems to be trying to push homosexuals down in a similar way. Liberalism and the ethic of autonomy are great protectors against such injustices. I believe it is dangerous for the ethic of divinity to supersede the ethic of autonomy in the governance of a diverse modern democracy. However, I also believe that life in a society that entirely ignored the ethic of divinity would be ugly and unsatisfying.

Because the culture war is ideological, both sides use the myth of pure evil. To acknowledge that the other side might be right about *anything* is an act of treason. My research on the third dimension, however, has freed me from the myth and made it easy for me to think treasonous thoughts. Here's one: If the third dimension and perceptions of sacredness are an important part of human nature, then the scientific community should accept religiosity as a normal and healthy aspect of human nature—an aspect that is as deep, important, and interesting as sexuality or language (which we study intensely). Here's another treasonous thought: If religious people are right in believing that religion is the source of their greatest happiness, then maybe the rest of us who are looking for happiness and meaning can learn something from them, whether or not we believe in God. That's the topic of the final chapter.

# 10

## Happiness
## Comes from Between

*Who sees all beings in his own Self, and his own Self in all*
*beings, loses all fear. . . . When a sage sees this great Unity*
*and his Self has become all beings, what delusion and what*
*sorrow can ever be near him?*

— Upanishads[1]

*I was entirely happy. Perhaps we feel like that when we die*
*and become a part of something entire, whether it is sun and*
*air, or goodness and knowledge. At any rate, that is happi-*
*ness: to be dissolved into something complete and great.*

— Willa Cather[2]

Proverbs, sayings, and words of wisdom dignify events, so we often use
them to mark important transitions in life. For the graduating class of 1981
at Scarsdale High School, in Scarsdale, New York, choosing a quotation
was a rite of passage, an opportunity to reflect on one's emerging identity
and express some aspect of it. As I look through the yearbook from that
class, at the quotations underneath each photo, I see two main kinds.
Many are tributes to love and friendship, appropriate for a time of parting
from friends ("You never really leave the friends you love. Part of them you

take with you, leaving a part of you behind." [ANONYMOUS]). The other kind expresses optimism, sometimes mixed with trepidation, about the road ahead. Indeed, it is difficult to think about graduating from high school without using the metaphor that life is a journey. For example, four students quoted the Cat Stevens song "On the Road to Find Out."[3] Two quoted George Washington: "I am embarked on a wide ocean, boundless in its prospect and, in which, perhaps, no safe harbor is to be found."[4] And one student quoted this line from Bruce Springsteen: "Well I got some beer and the highway's free / and I got you, and baby you've got me."[5]

But nestled among these affirmations of life's limitless possibilities is one with a darker tone: "Whosoever shall not fall by the sword or by famine, shall fall by pestilence so why bother shaving?" (WOODY ALLEN).[6] Above those words is a photograph of me.

I was only partly kidding. During the previous year, I had written a paper examining the play *Waiting for Godot*, Samuel Beckett's existentialist meditation on the absurdity of life in a world with no God, and it got me thinking. I was already an atheist, and by my senior year I had became obsessed with the question "What is the meaning of life?" I wrote my personal statement for college admissions on the meaninglessness of life. I spent the winter of my senior year in a kind of philosophical depression—not a clinical depression, just a pervasive sense that everything was pointless. In the grand scheme of things, I thought, it really didn't matter whether I got into college, or whether the Earth was destroyed by an asteroid or by nuclear war.

My despair was particularly strange because, for the first time since the age of four, my life was perfect. I had a wonderful girlfriend, great friends, and loving parents. I was captain of the track team, and, perhaps most important for a seventeen-year-old boy, I got to drive around in my father's 1966 Thunderbird convertible. Yet I kept wondering why any of it mattered. Like the author of Ecclesiastes, I thought that "all is vanity and a chasing after wind" (ECCLESIASTES 1:14).

I finally escaped when, after a week of thinking about suicide (in the abstract, not as a plan), I turned the problem inside out. There is no God and no externally given meaning to life, I thought, so from one perspective it really wouldn't matter if I killed myself tomorrow. Very well, then everything beyond tomorrow is a gift with no strings and no expectations. There is no test to hand in at the end of life, so there is no way to fail. If this really is all there is, why not embrace it, rather than throw it away? I don't know whether this realization lifted my mood or whether an improving mood helped me to reframe the problem with hope; but my existential depression lifted and I enjoyed the last months of high school.

My interest in the meaning of life continued, however, so in college I majored in philosophy, where I found few answers. Modern philosophers specialize in analyzing the meaning of words, but, aside from the existentialists (who caused the problem for me in the first place), they had little to say about the meaning of life. It was only after I entered graduate school in psychology that I realized why modern philosophy seemed sterile: It lacked a deep understanding of human nature. The ancient philosophers were often good psychologists, as I have shown in this book, but when modern philosophy began to devote itself to the study of logic and rationality, it gradually lost interest in psychology and lost touch with the passionate, contextualized nature of human life. It is impossible to analyze "the meaning of life" in the abstract, or in general, or for some mythical and perfectly rational being.[7] Only by knowing the kinds of beings that we actually are, with the complex mental and emotional architecture that we happen to possess, can anyone even begin to ask about what would count as a meaningful life. (Philosophy has, to its credit, become more psychological and more passionate in recent years.)[8]

As I went on in psychology and in my own research on morality, I discovered that psychology and related sciences have revealed so much about human nature that an answer is now possible. In fact, we've known most of the answer for a hundred years, and many of the remaining pieces have fallen into place over the last ten. This chapter is my version of psychology's answer to the ultimate question.

## WHAT WAS THE QUESTION?

The question "What is the meaning of life?" might be called the Holy Question, in analogy to the Holy Grail: Its pursuit is noble and everyone should want to find an answer, yet few people expect that one can be found. That's why books and movies that purport to tell us the answer to the Holy Question often do so only in jest. In *The Hitchhiker's Guide to the Galaxy,* a gigantic computer built to answer the Holy Question spits out its solution after 7.5 million years of computation: "forty-two."[9] In the closing scene of the movie *Monty Python's The Meaning of Life,* the answer to the Holy Question is handed to the actor Michael Palin (in drag), who reads it aloud: "Try to be nice to people, avoid eating fat, read a good book every now and then, get some walking in, and try to live in harmony with people of all creeds and nations."[10] These answers are funny precisely because they take the *form* of good answers, yet their *content* is empty or mundane. These parodies invite us to laugh at ourselves and ask: What was I expecting? What kind of answer *could* have satisfied me?

One thing philosophy did teach me is how to analyze questions, how to clarify exactly what is being asked before giving an answer. The Holy Question cries out for clarification. Whenever we ask "What is the meaning of X?" what *kind* of answer could possibly satisfy us?

The most common kind of meaning is definitional. "What is the meaning of 'ananym'?" means "Define the word 'ananym' for me so that I can understand it when I read it." I go to a dictionary,[11] look it up, and find that it means "a pseudonym consisting of the real name written backwards." Very well, what is the meaning of "life"? I go back to the dictionary and find that life has twenty-one meanings, including "the quality that distinguishes a vital and functional being from a dead body or purely chemical matter" and "the period from birth to death." Dead end. This is not at all the right kind of answer. We are not asking about the *word* "life," we're asking about life itself.

A second kind of meaning is about symbolism or substitution. If you dream about exploring a basement and finding a trap door to a subbasement, you might ask, "What is the meaning of the subbasement?" The psychologist Carl Jung had such a dream[12] and concluded that the meaning of the subbasement—the thing it symbolized or stood for—was the collective unconscious, a deep set of ideas shared by all people. But this is another dead end. Life does not symbolize, stand for, or point to anything. It is life itself that we want to understand.

A third way in which we ask about meaning is as a plea for help in making sense of something, usually with reference to people's intentions and beliefs. Suppose you walk into a movie half an hour late and have to leave half an hour before the end. Later that night you are talking with a friend who saw the whole film and you ask, "What did it mean when the guy with the curly hair winked at that kid?" You are aware that the act had some significance for the plot of the movie, and you suspect that you need to know certain facts to understand that act. Perhaps a prior relationship between the two characters had been revealed in the opening scenes? To ask, "What was the meaning of the wink?" really means, "What do I need to know to understand that wink?" Now we're making progress, for life is much like a movie we walk into well after its opening scene, and we will have to step out long before most of the story lines reach their conclusions. We are acutely aware that we need to know a great deal if we are to understand the few confusing minutes that we do watch. Of course, we don't know exactly what it is that we don't know, so we can't frame the question well. We ask, "What is the meaning of life?" not expecting a direct answer (such as "forty-two"), but rather hoping for some enlightenment, something to give us an "aha!" experience in which, suddenly, things that we had not before understood or recognized as important begin to make sense (as they did for the square taken to the third dimension).

Once the Holy Question has been re-framed to mean "Tell me something enlightening about life," the answer must involve the kinds of revelations that human beings find enlightening. There appear to be two specific sub-questions to which people want answers, and for which they find answers enlightening. The first can be called the question of the purpose *of* life: "What is the purpose *for which* human beings were placed on Earth? *Why* are we here?" There are two major classes of answers to this question: Either you believe in a god/spirit/intelligence who had some idea, desire, or intention in creating the world or you believe in a purely material world in which it and you were not created *for* any reason; it all just happened as matter and energy interacted according to the laws of nature (which, once life got started, included the principles of Darwinian evolution). Religion is often seen as an answer to the Holy Question because many religions offer such clear answers to the sub-question of the purpose *of* life. Science and religion are often seen as antagonists, and, indeed, they battle over the teaching of evolution in the United States precisely because they offer conflicting answers.

The second sub-question is the question of purpose *within* life: "How ought I to live? What should I do to have a good, happy, fulfilling, and *meaningful* life?" When people ask the Holy Question, one of the things they are hoping for is a set of principles or goals that can guide their actions and give their choices meaning or value. (That is why the form of the answer in the Monty Python movie is correct: "Try to be nice to people, avoid eating fat . . . "). Aristotle asked about *aretē* (excellence/virtue) and *telos* (purpose/goal), and he used the metaphor that people are like archers, who need a clear target at which to aim.[13] Without a target or goal, one is left with the animal default: Just let the elephant graze or roam where he pleases. And because elephants live in herds, one ends up doing what everyone else is doing. Yet the human mind has a rider, and as the rider begins to think more abstractly in adolescence, there may come a time when he looks around, past the edges of the herd, and asks: Where are we all going? And why? This is what happened to me my senior year of high school.

In my adolescent existentialism, I conflated the two sub-questions. Because I embraced the scientific answer to the question of the purpose *of* life, I thought it precluded finding purpose *within* life. It was an easy mistake to make because many religions teach that the two questions are inseparable. If you believe that God created you as part of His plan, then you can figure out how you ought to live if you are going to play your part properly. *The Purpose Driven Life*[14] is a forty-day course that teaches readers how to find purpose *within* life from the theological answer to the question of the purpose *of* life.

The two questions can, however, be separated. The first asks about life from the outside; it looks at people, the Earth, and the stars as *objects*—"Why do *they* all exist?"—and is properly addressed by theologians, physicists, and biologists. The second question is about life from the inside, as a *subject*—"How can *I* find a sense of meaning and purpose?"—and is properly addressed by theologians, philosophers, and psychologists. The second question is really empirical—a question of fact that can be examined by scientific means. Why do some people live lives full of zest, commitment, and meaning, but others feel that their lives are empty and pointless? For the rest of this chapter I will ignore the purpose *of* life and search for the factors that give rise to a sense of purpose *within* life.

## LOVE AND WORK

When a computer breaks, it doesn't fix itself. You have to open it up and do something to it, or bring it to a specialist for repair. The computer metaphor has so pervaded our thought that we sometimes think about people as computers, and about psychotherapy as the repair shop or a kind of reprogramming. But people are not computers, and they usually recover on their own from almost anything that happens to them.[15] I think a better metaphor is that people are like plants. During graduate school, I had a small garden in front of my house in Philadelphia. I was not a very good gardener, and I traveled a lot in the summers, so sometimes my plants withered and nearly died. But the amazing thing I learned about plants is that as long as they are not completely dead, they will spring back to full and glorious life if you just get the conditions right. You can't fix a plant; you can only give it the right conditions—water, sun, and soil—and then wait. It will do the rest.

If people are like plants, what are the conditions we need to flourish? In the happiness formula from chapter 5, H(appiness) = S(etpoint) + C(onditions) + V(oluntary activities), what exactly is C? The biggest part of C, as I said in chapter 6, is love. No man, woman, or child is an island. We are ultrasocial creatures, and we can't be happy without having friends and secure attachments to other people. The second most important part of C is having and pursuing the right goals, in order to create states of flow and engagement. In the modern world, people can find goals and flow in many settings, but most people find most of their flow at work.[16] (I define work broadly to include anyone's answer to the question "So, what do you do?" "Student" and "full-time parent" are both good answers). Love and work are, for people, obvious analogues to water and sunshine for plants.[17] When Freud was asked what a normal person should be able to do well, he is reputed to have said, "Love and work."[18] If therapy can help a person do those two things well, it has succeeded. In Maslow's famous hierarchy of needs, once people have satisfied their physical needs (such as food and safety), they move on to needs for love and then esteem, which is earned mostly through one's work. Even before Freud, Leo Tolstoy wrote: "One can live magnificently in this world, if one knows how to work and how to love, to work for the person one loves and to love one's work."[19] Having earlier said everything I want to say about love, I will say no more here. But I must say much more about work.

When Harry Harlow took his students to the zoo, they were surprised to find that apes and monkeys would solve problems just for the fun of it. Behaviorism had no way to explain such unreinforced behavior. In 1959, the Harvard psychologist Robert White[20] concluded, after surveying research in behaviorism and psychoanalysis, that both theories had missed what Harlow had noticed: the overwhelming evidence that people and many other mammals have a basic drive to *make things happen*. You can see it in the joy infants take with "busy boxes," the activity centers that allow them to convert flailing arm movements into ringing bells and spinning wheels. You can see it in the toys to which older children gravitate. The ones I most intensely longed for as a boy were those that caused movement or action at a distance: remote-controlled cars, guns that shot plastic pellets, and rockets or airplanes of any kind. And you can see it in the lethargy that often overtakes people who stop working, whether from retirement, being fired, or winning a lottery. Psychologists have referred to this basic need as a need for competence, industry, or mastery. White called it the "effectance motive," which he defined as the need or drive to develop competence through interacting with and controlling one's environment. Effectance is almost as basic a need as food and water, yet it is not a deficit need, like hunger, that is satisfied and then disappears for a few hours. Rather, White said, effectance is a constant presence in our lives:

> Dealing with the environment means carrying on a continuing transaction which gradually changes one's relation to the environment. Because there is no consummatory climax, satisfaction has to be seen as lying in a considerable series of transactions, in a trend of behavior rather than a goal that is achieved.[21]

The effectance motive helps explain the progress principle: We get more pleasure from making progress toward our goals than we do from achieving them because, as Shakespeare said, "Joy's soul lies in the doing."[22]

Now we can look at the conditions of modern work. Karl Marx's criticism of capitalism[23] was based in part on his justified claim that the Industrial Revolution had destroyed the historical relationship between craftsmen and the goods they produced. Assembly-line work turned people into cogs in a giant machine, and the machine didn't care about workers' need for effectance. Later research on job satisfaction supported Marx's critique, but added nuance. In 1964, the sociologists Melvin Kohn and Carmi Schooler[24] surveyed 3,100 American men about their jobs and found that the key to understanding which jobs were satisfying was what they called "occupational self direction." Men who were closely supervised in jobs of low complexity and much routine showed the highest degree of alienation (feeling powerless, dissatisfied, and separated from the work). Men who had more latitude in deciding how they approached work that was varied and challenging tended to enjoy their work much more. When workers had occupational self-direction, their work was often satisfying.

More recent research finds that most people approach their work in one of three ways: as a job, a career, or a calling.[25] If you see your work as a job, you do it only for the money, you look at the clock frequently while dreaming about the weekend ahead, and you probably pursue hobbies, which satisfy your effectance needs more thoroughly than does your work. If you see your work as a career, you have larger goals of advancement, promotion, and prestige. The pursuit of these goals often energizes you, and you sometimes take work home with you because you want to get the job done properly. Yet, at times, you wonder why you work so hard. You might occasionally see your work as a rat race where people are competing for the sake of competing. If you see your work as a calling, however, you find your work intrinsically fulfilling—you are not doing it to achieve something else. You see your work as contributing to the greater good or as playing a role in some larger enterprise the worth of which seems obvious to you. You have frequent experiences of flow during the work day, and you neither look forward to "quitting time" nor feel the desire to shout, "Thank God it's Friday!" You would continue to work, perhaps even without pay, if you suddenly became very wealthy.

You might think that blue-collar workers have jobs, managers have careers, and the more respected professionals (doctors, scientists, clergy) have callings. Although there is some truth to that expectation, we can nonetheless paraphrase Marcus Aurelius and say, "Work itself is but what you deem it." Amy Wrzesniewski, a psychologist at New York University, finds all three orientations represented in almost every occupation she has examined.[26] In a study of hospital workers, for example, she found that the janitors who cleaned bed pans and mopped up vomit—perhaps the lowest-ranking job in a hospital—sometimes saw themselves as part of a team whose goal was to heal people. They went beyond the minimum requirements of their job description, for example, by trying to brighten up the rooms of very sick patients or anticipating the needs of the doctors and nurses rather than waiting for orders. In so doing, they increased their own occupational self-direction and created for themselves jobs that satisfied their effectance needs. Those janitors who worked this way saw their work as a calling and enjoyed it far more than those who saw it as a job.

The optimistic conclusion coming out of research in positive psychology is that most people can get more satisfaction from their work. The first step is to know your strengths. Take the strengths test[27] and then choose work that allows you to use your strengths every day, thereby giving yourself at least scattered moments of flow. If you are stuck in a job that doesn't match your strengths, recast and reframe your job so that it does. Maybe you'll have to do some extra work for a while, like the hospital janitors who were acting on strengths of kindness, loving, emotional intelligence, or citizenship. If you *can* engage your strengths, you'll find more gratification in work; if you find gratification, you'll shift into a more positive, approach-oriented mindset; and in such a mindset it will be easier for you to see the bigger picture[28]—the contribution you are making to a larger enterprise—within which your job might turn into a calling.

Work at its best, then, is about connection, engagement, and commitment. As the poet Kahlil Gibran said, "Work is love made visible." Echoing Tolstoy, he gave examples of work done with love:

> It is to weave the cloth with threads drawn from your heart,
> even as if your beloved were to wear that cloth.

> It is to build a house with affection,
> even as if your beloved were to dwell in that house.
> It is to sow seeds with tenderness and reap the harvest with joy,
> even as if your beloved were to eat the fruit.[29]

Love and work are crucial for human happiness because, when done well, they draw us out of ourselves and into connection with people and projects beyond ourselves. Happiness comes from getting these connections right. Happiness comes not just from within, as Buddha and Epictetus supposed, or even from a combination of internal and external factors (as I suggested as a temporary fix at the end of chapter 5). The correct version of the happiness hypothesis, as I'll illustrate below, is that happiness comes from *between*.

## VITAL ENGAGEMENT

Plants thrive under particular conditions, and biologists can now tell us how sunlight and water get converted into plant growth. People thrive under particular conditions, and psychologists can now tell us how love and work get converted into happiness and a sense of meaning.

The man who found flow, Mihalyi Csikszentmihalyi, thinks big. Not content to study moments of flow (by beeping people several times a day), he wanted to know what role flow plays in life as a whole, particularly in the lives of creative people. So he turned to the experts: paragons of success in the arts and sciences. He and his students have interviewed hundreds of successful painters, dancers, poets, novelists, physicists, biologists, and psychologists—all people who seem to have crafted lives for themselves built around a consuming passion. These are admirable lives, desirable lives, the sort that many young people dream of having when they look to these people as role models. Csikszentmihalyi wanted to know how such lives happened. How does a person come to make such a commitment to a field and then become so extraordinarily creative?

His interviews showed that every path is unique, yet most of them led in the same direction: from initial interest and enjoyment, with moments of flow, through a relationship to people, practices, and values that deepened over many years, thereby enabling even longer periods of flow. Csikszent-mihalyi and his students, particularly Jeanne Nakamura, have studied the end state of this deepening process and called it "vital engagement," which they define as "a relationship to the world that is characterized both by experiences of flow (enjoyed absorption) and by meaning (subjective significance)."[30] Vital engagement is another way of saying that work has become "love made visible"; Nakamura and Csikszentmihalyi even describe vital engagement in words that could almost have been taken from a romance novel: "There is a strong felt connection between self and object; a writer is 'swept away' by a project, a scientist is 'mesmerized by the stars.' The relationship has subjective meaning; work is a 'calling.'"[31]

Vital engagement is a subtle concept, and the first time I taught a course on positive psychology, the students weren't getting it. I thought that an example would help, so I called on a woman who had been quiet in class, but who had once mentioned her interest in horses. I asked Katherine to tell us how she got involved in riding. She described her childhood love of animals, and her interest in horses in particular. At the age of ten she begged her parents to let her take riding lessons, and they agreed. She rode for fun at first, but soon began riding in competitions. When it came time to choose a college, she chose the University of Virginia in part because it had an excellent riding team.

Katherine was shy, and, after narrating these basic facts, she stopped talking. She had told us about her increasing commitment to riding, but vital engagement is more than just commitment. I probed further. I asked whether she could tell us the names of specific horses from previous centuries. She smiled and said, almost as if admitting a secret, that she had begun to read about horses when she began to ride, and that she knew a great deal about the history of horses and about famous horses in history. I asked whether she had made friends through riding, and she told us that most of her close friends were "horse friends," people she knew from horse shows and from riding together. As she talked, she grew more animated and confident. It was as clear from her demeanor as from her words that Katherine had found vital engagement in riding. Just as Nakamura and Csikszentmihalyi had said, her initial interest grew into an ever-deepening

relationship, an ever-thickening web connecting her to an activity, a tradition, and a community. Riding for Katherine had become a source of flow,
joy, identity, effectance, and relatedness. It was part of her answer to the
question of purpose within life.

Vital engagement does not reside in the person or in the environment; it
exists in the relationship *between* the two. The web of meaning that engulfed
Katherine grew and thickened gradually and organically, over many years. Vital engagement is what I was missing during my senior year of high school. I
had love, and I had work (in the form of reasonably challenging high school
classes), but my work was not part of a larger project beyond getting into college. In fact, it was precisely when the college project was ending—when I
had sent off my college applications and was in limbo, not knowing where
I would go next—that I became paralyzed by the Holy Question.

Getting the right relationship between you and your work is not entirely
up to you. Some occupations come ready-made for vital engagement; others
make it difficult. As market forces were reshaping many professions in the
United States during the 1990s—medicine, journalism, science, education, and the arts—people in those fields began to complain that the quality of work and the quality of life were sometimes compromised by the
relentless drive to increase profits. Csikszentmihalyi teamed up with two
other leading psychologists—Howard Gardner at Harvard, and William Damon at Stanford—to study these changes, and to see why some professions
seemed healthy while others were growing sick. Picking the fields of genetics and journalism as case studies, they conducted dozens of interviews
with people in each field. Their conclusion[32] is as profound as it is simple:
It's a matter of alignment. When doing *good* (doing high-quality work that
produces something of use to others) matches up with doing *well* (achieving wealth and professional advancement), a field is healthy. Genetics, for
example, is a healthy field because all parties involved respect and reward
the very best science. Even though pharmaceutical companies and market
forces were beginning to inject vast amounts of money into university research labs in the 1990s, the scientists whom Csikszentmihalyi, Gardner,
and Damon interviewed did not believe they were being asked to lower
their standards, cheat, lie, or sell their souls. Geneticists believed that their
field was in a golden age in which excellent work brought great benefits to

the general public, the pharmaceutical companies, the universities, and the scientists themselves.

Journalists, on the other hand, were in trouble. Most of them had gone into journalism with high ideals—respect for the truth, a desire to make a difference in the world, and a firm belief that a free press is a crucial support of democracy. But by the 1990s, the decline of family-run newspapers and the rise of corporate media empires had converted American journalism into just another profit center where the only thing that mattered was will it sell, and will it outsell our competitors? Good journalism was sometimes bad for business. Scare stories, exaggeration, trumped up conflict, and sexual scandal, all cut up into tiny digestible pieces, were often more profitable. Many journalists who worked for these empires confessed to having a sense of being forced to sell out and violate their own moral standards. Their world was unaligned, and they could not become vitally engaged in the larger but ignoble mission of gaining market share at any cost.

## CROSS-LEVEL COHERENCE

The word "coherence" literally means holding or sticking together, but it is usually used to refer to a system, an idea, or a worldview whose parts fit together in a consistent and efficient way. Coherent things work well: A coherent worldview can explain almost anything, while an incoherent worldview is hobbled by internal contradictions. A coherent profession, such as genetics, can get on with the business of genetics, while an incoherent profession, like journalism, spends a lot of time on self-analysis and self-criticism.[33] Most people know there's a problem, but they can't agree on what to do about it.

Whenever a system can be analyzed at multiple levels, a special kind of coherence occurs when the levels mesh and mutually interlock. We saw this cross-level coherence in the analysis of personality: If your lower-level traits match up with your coping mechanisms, which in turn are consistent with your life story, your personality is well integrated and you can get on with the business of living. When these levels do not cohere, you are likely to be torn by internal contradictions and neurotic conflicts.[34] You might need adversity to knock yourself into alignment. And if you do achieve coherence, the moment when things come together may be one of the most profound of your life. Like the moviegoer who later finds out what she missed in the first half hour, your life will suddenly make more sense. Finding coherence across levels feels like enlightenment,[35] and it is crucial for answering the question of purpose within life.

People are multilevel systems in another way: We are *physical* objects (bodies and brains) from which *minds* somehow emerge; and from our minds, somehow *societies* and *cultures* form.[36] To understand ourselves fully we must study all three levels—physical, psychological, and sociocultural. There has long been a division of academic labor: Biologists studied the brain as a physical object, psychologists studied the mind, and sociologists and anthropologists studied the socially constructed environments within which minds develop and function. But a division of labor is productive only when the tasks are coherent—when all lines of work eventually combine to make something greater than the sum of its parts. For much of the twentieth century that didn't happen—each field ignored the others and focused on its own questions. But nowadays cross-disciplinary work is flourishing, spreading out from the middle level (psychology) along bridges (or perhaps ladders) down to the physical level (for example, the field of cognitive neuroscience) and up to the sociocultural level (for example, cultural psychology). The sciences are linking up, generating cross-level coherence, and, like magic, big new ideas are beginning to emerge.

Here is one of the most profound ideas to come from the ongoing synthesis: *People gain a sense of meaning when their lives cohere across the three levels of their existence.*[37] The best way I can illustrate this idea is to take you back to Bhubaneswar, India. I have already explained the logic of purity and pollution, so you understand why Hindus bathe before making an offering to God, and why they are careful about what they touch on the way to the temple. You understand why contact with a dog, a menstruating woman, or a person of low caste can render a person of high caste temporarily impure and unfit to make an offering. But you understand all this only at the psychological level and, even then, only as a set of propositions grasped by the rider and stored away as explicit knowledge. You do not feel polluted after touching the arm of a woman you know to be menstruating; you do not even know what it would feel like to feel polluted in that way.

Suppose, however, that you grow up as a Brahmin in Bhubaneswar. Every day of your life you have to respect the invisible lines separating pure from profane spaces, and you have to keep track of people's fluctuating levels of purity before you can touch them or take anything from their hands. You bathe several times a day—short baths or brief immersions in sacred water—always before making a religious offering. And your offerings are not just words: You actually give some food to God (the priest touches your offering to the image, icon, or object in the inner sanctum), which is returned to you so that you may eat what God left over. Eating someone's leftovers shows a willingness to take in that person's saliva, which demonstrates both intimacy and subordination in Bhubaneswar. Eating God's leftovers is an act of intimacy, and subordination, too. After twenty years of these practices, your understanding of Hindu rituals is *visceral*. Your explicit understanding is supported by a hundred physical feelings: shivering during the morning bath at sunrise; the pleasure of washing off dust and putting on clean clothes after a bath on a hot afternoon; the feeling of bare feet on cool stone floors as you approach the inner sanctum; the smell of incense; the sound of mumbled prayers in Sanskrit, the bland (pure) taste of rice that has been returned to you from God. In all these ways, your understanding at the psychological level has spread down to your physical embodiment, and when the conceptual and visceral levels connect, the rituals *feel* right to you.

Your understanding of ritual spreads up to the sociocultural level, too. You are immersed in a 4,000-year-old religious tradition that provided most of the stories you heard as a child, many of which involved plot elements of purity and pollution. Hinduism structures your social space through a caste system based on the purity and pollution of various occupations, and it structures your physical space with the topography of purity and pollution that keeps temples, kitchens, and right hands pure. Hinduism also gives you a cosmology in which souls reincarnate by moving up or down on the vertical dimension of divinity. So every time you make an offering to God, the three levels of your existence are all aligned and mutually interlocking. Your physical feelings and conscious thoughts cohere with your actions, and all of it makes perfect sense within the larger culture of which you are a part. As you make an offering to God, you don't think, "What does this all mean? Why am I doing this?" The experience of meaningfulness just happens. It emerges automatically from cross-level coherence. Once again, happiness—or a sense of meaningfulness that imparts richness to experience—comes from between.

In contrast, think about the last empty ritual you took part in. Maybe you were asked to join hands and chant with a group of strangers while attending a wedding ceremony for a friend who is of a different religion. Perhaps you took part in a new age ceremony that borrowed elements from Native Americans, ancient Celts, and Tibetan Buddhists. You probably understood the symbolism of the ritual—understood it consciously and explicitly in the way that the rider is so good at doing. Yet you felt self-conscious, maybe even silly, while doing it. Something was missing.

You can't just invent a good ritual through reasoning about symbolism. You need a tradition within which the symbols are embedded, and you need to invoke bodily feelings that have some appropriate associations. Then you need a community to endorse and practice it over time. To the extent that a community has many rituals that cohere across the three levels, people in the community are likely to feel themselves connected to the community and its traditions. If the community also offers guidance on how to live and what is of value, then people are unlikely to wonder about the question of purpose within life. Meaning and purpose simply emerge from the coherence, and people can get on with the business of living. But conflict, paralysis, and anomie are likely when a community fails to provide coherence, or, worse, when its practices contradict people's gut feelings or their shared mythology and ideology. (Martin Luther King, Jr., forced Americans to confront contradictions between practices of racial segregation and ideals about equality and freedom. Many people didn't like that.) People don't necessarily need to find meaning in their national identity—indeed, in large and diverse nations such as the United States, Russia, and India, religion might hold greater promise for cross-level coherence and purpose within life. Religions do such a good job of creating coherence, in fact, that some scholars[38] believe they were designed for that purpose.

## GOD GIVES US HIVES

When I first began to study morality as a philosophy major in college, my father said, "Why aren't you studying religion, too? How could people have morality without God?" As a young atheist with a strong sense of morality (well over the border into self-righteousness), I was insulted by my father's suggestion. Morality, I thought, was about relationships among people; it was about a commitment to doing the right thing, even when it goes against your self-interest. Religion, I thought, was a bunch of rules that made no sense and stories that could never have happened, written down by people and then falsely attributed to a supernatural entity.

I now believe my father was right—morality has its origins in religion—but not for the reasons he believed. Morality and religion both occur in some form in all human cultures[39] and are almost always both intertwined with the values, identity, and daily life of the culture. Anyone who wants a full, cross-level account of human nature, and of how human beings find purpose and meaning in their lives, must make that account cohere with what is known about morality and religion.

From an evolutionary perspective, morality is a problem. If evolution is all about survival of the fittest, then why do people help each other so much? Why do they give to charity, risk their lives to save strangers, and volunteer to fight in wars? Darwin thought the answer was easy: Altruism evolves for the good of the group:

> There can be no doubt that a tribe including many members who, from possessing in a high degree the spirit of patriotism, fidelity, obedience, courage, and sympathy, were always ready to aid one another, and to sacrifice themselves for the common good would be victorious over most other tribes, and this would be natural selection.[40]

Darwin proposed that groups compete, just like individuals, and therefore psychological features that make groups successful—such as patriotism, courage, and altruism toward fellow group members—should spread like any other trait. But once evolutionary theorists began testing predictions rigorously, using computers to model the interactions of individuals who use various strategies (such as pure selfishness versus tit for tat), they quickly came to appreciate the seriousness of the "free-rider problem." In groups in which people make sacrifices for the common good, an individual who makes no such sacrifices—who in effect takes a free ride on the backs of the altruists—comes out ahead. In the cold logic of these computer simulations, whoever accumulates the most resources in one generation goes on to produce more children in the next, so selfishness is adaptive but altruism is not. The only solution to the free-rider problem is to make altruism pay, and two back-to-back breakthroughs in evolutionary thinking showed how to do that. In chapter 3 I presented kin altruism (be nice to those who share your genes) and reciprocal altruism (be nice to those who might reciprocate in the future) as two steps on the way to ultrasociality. Once these two solutions to the free-rider problem were published (in 1966 and 1971, respectively),[41] most evolutionary theorists considered the problem of altruism solved and essentially declared group selection illegal. Altruism could be explained away as a special kind of selfishness, and anyone who followed Darwin in thinking that evolution worked for the "good of the group" instead of the good of the individual (or better yet, the good of the gene),[42] was dismissed as a mushy-headed romantic.

The ban on group selection had one loophole. For creatures that really do compete, live, and die as a group, such as the other ultrasocial animals (bees, wasps, ants, termites, and naked mole rats), group selection explanations were appropriate. There is a real sense in which a beehive or an ant colony is a single organism, each insect a cell in the larger body.[43] Like stem cells, ants can take different physical forms to perform specific functions needed by the colony: small bodies to care for larva, larger bodies with special appendages to forage for food or fight off attackers. Like cells in the immune system, ants will sacrifice themselves to protect the colony: In one species of Malaysian ant,[44] members of the soldier caste store a sticky substance just under their exoskeletons. In the midst of battle, they explode their bodies, turning themselves into suicide bombers to gum up their adversaries. For ants and bees, the queen is not the brain; she is the ovary, and the entire hive or colony can be seen as a body shaped by natural selection to protect the ovary and help it create more hives or colonies. Because all members really are in the same boat, group selection is not just permissible as an explanation; it is mandatory.

Might this loophole apply to humans as well? Do humans compete, live, and die as a group? Tribes and ethnic groups do grow and spread or fade and die out, and sometimes this process has occurred by genocide. Furthermore, human societies often have an extraordinary division of labor, so the comparison to bees and ants is tempting. But as long as each human being has the opportunity to reproduce, the evolutionary payoffs for investing in one's own welfare and one's own offspring will almost always exceed the payoffs for contributing to the group; in the long run, selfish traits will therefore spread at the expense of altruistic traits. Even during war and genocide, when group interests are most compelling, it is the coward who runs and hides, rather than joining his comrades on the front lines, who is most likely to pass on his genes to the next generation. Evolutionary theorists have therefore stood united, since the early 1970s, in their belief that group selection simply did not play a role in shaping human nature.

But wait a second. This is not an all-or-nothing issue. Even if the competition of individuals *within* a group is the most important process in human evolution, group selection (competition *between* groups) could have played a role too. The evolutionary biologist David Sloan Wilson[45] has recently argued that the banishment of group selection theories on the basis of some oversimplified computer models from the 1960s was one of the biggest mistakes in the history of modern biology. If you make the models more realistic, more like real human beings, group selection jumps right out at you. Wilson points out that human beings evolve at two levels simultaneously: genetic and cultural. The simple models of the 1960s worked well for creatures without culture; for them, behavioral traits must all be encoded in the genes, which are passed on only along lines of kinship. But everything a person does is influenced not only by her genes but also by her culture, and cultures evolve, too. Because elements of culture show variation (people invent new things) and selection (other people do or don't adopt those variations), cultural traits can be analyzed in a Darwinian framework[46] just as well as physical traits (birds' beaks, giraffes' necks). Cultural elements, however, don't spread by the slow process of having children; they spread rapidly whenever people adopt a new behavior, technology, or belief. Cultural traits can even spread from tribe to tribe or nation to nation, as when the plough, the printing press, or reality television programs became popular in many places in quick succession.

Cultural and genetic evolution are intertwined. The human capacity for culture—a strong tendency to learn from each other, to teach each other, and to build upon what we have learned—is itself a genetic innovation that happened in stages over the last few million years.[47] But once our brains reached a critical threshold, perhaps 80,000 to 100,000 years ago,[48] cultural innovation began to accelerate; a strong evolutionary pressure then shaped brains to take further advantage of culture. Individuals who could best learn from others were more successful than their less "cultured" brethren, and as brains became more cultural, cultures became more elaborate, further increasing the advantage of having a more cultural brain. All human beings today are the products of the co-evolution of a set of genes (which is almost identical across cultures) and a set of cultural elements (which is diverse across cultures, but still constrained by the capacities and predispositions of the human mind).[49] For example, the genetic evolution of the emotion of disgust made it possible (but not inevitable) for cultures to develop caste systems based on occupation and supported by disgust toward those who perform "polluting" activities. A caste system then restricts marriage to within-caste pairings, which in turn alters the course of genetic evolution. After a thousand years of inbreeding within caste, castes will diverge slightly on a few genetic traits—for example, shades of skin color—which might in turn lead to a growing cultural association of caste with color rather than just with occupation. (It only takes twenty generations of selective breeding to create large differences of appearance and behavior in other mammals.)[50] In this way, genes and cultures co-evolve;[51] they mutually affect each other, and neither process can be studied in isolation for human beings.

Wilson examines religion from this co-evolutionary perspective. The word *religion* literally means, in Latin, to link or bind together; and despite the vast variation in the world's religions, Wilson shows that religions always serve to coordinate and orient people's behavior toward each other and toward the group as a whole, sometimes for the purpose of competing with other groups. The sociologist Emile Durkheim first developed this view of religion in 1912:

> A religion is a unified system of beliefs and practices relative to sacred things, that is to say, things set apart and forbidden—beliefs and practices which unite into one single moral community called a church, all those who adhere to them.[52]

Wilson shows how religious practices help members solve coordination problems. For example, trust and therefore trade are greatly enhanced when all parties are part of the same religious community, and when religious beliefs say that God knows and cares about the honesty of the parties. (The anthropologist Pascal Boyer[53] points out that gods and ancestor spirits are often thought to be omniscient, yet what they most care about in this vast universe is the moral intentions hidden in the hearts of the living.) Respect for rules is enhanced when rules have an element of sacredness, and when they are backed up by supernatural sanction and the gossip or ostracism of one's peers. Wilson's claim is that religious ideas, and brains that responded to those ideas, co-evolved. Even if the belief in supernatural entities emerged originally for some other reason, or as an accidental byproduct in the evolution of cognition (as some scholars have claimed),[54] groups that parlayed those beliefs into social coordination devices (for example, by linking them to emotions such as shame, fear, guilt, and love) found a cultural solution to the free-rider problem and then reaped the enormous benefits of trust and cooperation. If stronger belief led to greater individual benefits, or if a group developed a way to punish or exclude those who did not share in its beliefs and practices, conditions were perfect for the co-evolution of religion and religious brains. (Consistent with Wilson's proposal, the geneticist Dean Hamer recently reported evidence from twin studies that suggests a particular gene may be associated with a stronger tendency to have religious and self-transcendent experiences.)[55]

Religion, therefore, could have pulled human beings into the group-selection loophole. By making people long ago feel and act as though they were part of one body, religion reduced the influence of individual selection (which shapes individuals to be selfish) and brought into play the force of group selection (which shapes individuals to work for the good of their group). But we didn't make it all the way through the loophole: Human nature is a complex mix of preparations for extreme selfishness and extreme altruism. Which side of our nature we express depends on culture and context. When opponents of evolution object that human beings are not mere apes, they are correct. We are also part bee.

## HARMONY AND PURPOSE

Reading Wilson's *Darwin's Cathedral* is like taking a journey to Spaceland. You can look down on the vast tapestry of human cultures and see why things are woven in the way that they are. Wilson says his own private hell would be to be locked forever into a room full of people discussing the hypocrisies of religion, for example, that many religions preach love, compassion, and virtue yet sometimes cause war, hatred, and terrorism. From Wilson's higher perspective, there is no contradiction. Group selection creates interlocking genetic and cultural adaptations that enhance peace, harmony, and cooperation *within* the group for the express purpose of increasing the group's ability to compete with *other* groups. Group selection does not end conflict; it just pushes it up to the next level of social organization. Atrocities committed in the name of religion are almost always committed against out-group members, or against the most dangerous people of all: apostates (who try to leave the group) and traitors (who undermine the group).

A second puzzle that Wilson can solve is why mysticism, everywhere and always, is about transcending the self and merging with something larger than the self. When William James analyzed mysticism, he focused on the psychological state of "cosmic consciousness"[56] and on the techniques developed in all the major religions to attain it. Hindus and Buddhists use meditation and yoga to attain the state of *samadhi*, in which "the subject-object distinction and one's sense of an individual self disappear in a state usually described as one of supreme peace, bliss, and illumination."[57] James found much the same goal in Christian and Muslim mysticism, often attained through repetitive prayer. He quoted the eleventh-century Muslim philosopher Al Ghazzali, who spent several years worshipping with the Sufis of Syria. Al Ghazzali attained experiences of "transport" and revelation that he said cannot be described in words, although he did try to explain to his Muslim readers the essence of Sufism:

> The first condition for a Sufi is to purge his heart entirely of all that is not God. The next key of the contemplative life consists in the humble prayers which escape from the fervent soul, and in the meditations on God in which the heart is swallowed up entirely. But in reality this is only the beginning of the Sufi life, the end of Sufism being total absorption in God.[58]

From Wilson's perspective, mystical experience is an "off" button for the self. When the self is turned off, people become just a cell in the larger body, a bee in the larger hive. It is no wonder that the after effects of mystical experience are predictable; people usually feel a stronger commitment to God or to helping others, often by bringing them to God.

The neuroscientist Andrew Newberg[59] has studied the brains of people undergoing mystical experiences, mostly during meditation, and has found where that off-switch might be. In the rear portion of the brain's parietal lobes (under the rear portion of the top of the skull) are two patches of cortex Newberg calls the "orientation association areas." The patch in the left hemisphere appears to contribute to the mental sensation of having a limited and physically defined body, and thus keeps track of your edges. The corresponding area in the right hemisphere maintains a map of the space around you. These two areas receive input from your senses to help them maintain an ongoing representation of your self and its location in space. At the very moment when people report achieving states of mystical union, these two areas appear to be cut off. Input from other parts of the brain is reduced, and overall activity in these orientation areas is reduced, too. But Newberg believes they are still trying to do their jobs: The area on the left tries to establish the body's boundaries and doesn't find them; the area on the right tries to establish the self's location in space and doesn't find it. The person experiences a loss of self combined with a paradoxical expansion of the self out into space, yet with no fixed location in the normal world of three dimensions. The person feels merged with something vast, something larger than the self.

Newberg believes that rituals that involve repetitive movement and chanting, particularly when they are performed by many people at the same time, help to set up "resonance patterns" in the brains of the participants that make this mystical state more likely to happen. The historian William McNeill, drawing on very different data, came to the same conclusion. When McNeill was drafted into the U.S. Army in 1941, basic training required that he march for hundreds of hours on the drill field in close formation with a few dozen other men. At first, McNeill thought the marching was just a way to pass the time because his base had no weapons with which to train. But after a few weeks of training, the marching began to induce in him an altered state of consciousness:

> Words are inadequate to describe the emotion aroused by the prolonged movement in unison that drilling involved. A sense of pervasive well-being is what I recall; more specifically, a strange sense of personal enlargement; a sort of swelling out, becoming bigger than life, thanks to participation in collective ritual.[60]

Decades later, McNeill studied the role that synchronized movement—in dance, religious ritual, and military training—has played in history. In *Keeping Together in Time,*[61] he concludes that human societies since the beginning of recorded history have used synchronized movement to create harmony and cohesion within groups, sometimes in the service of preparing for hostilities with other groups. McNeill's conclusion suggests that synchronized movement and chanting might be evolved mechanisms for activating the altruistic motivations created in the process of group selection. The extreme self-sacrifice characteristic of group-selected species such as ants and bees can often be found among soldiers. McNeill quotes an extraordinary passage from the book *The Warriors: Reflections of Men in Battle* that describes the thrilling communal state that soldiers sometimes enter:

> "I" passes insensibly into a "we," "my" becomes "our," and individual fate loses its central importance. . . . I believe that it is nothing less than the assurance of immortality that makes self-sacrifice at these moments so relatively easy. . . . I may fall, but I do not die, for that which is real in me goes forward and lives on in the comrades for whom I gave up my life.[62]

There is indeed something larger than the self, able to provide people with a sense of purpose they think worth dying for: the group. (Of course, one group's noble purpose is sometimes another group's pure evil.)

## THE MEANING OF LIFE

What can you do to have a good, happy, fulfilling, and meaningful life? What is the answer to the question of purpose *within* life? I believe the answer can be found only by understanding the kind of creature that we are, divided in the many ways we are divided. We were shaped by individual selection to be selfish creatures who struggle for resources, pleasure, and prestige, and we were shaped by group selection to be hive creatures who long to lose ourselves in something larger. We are social creatures who need love and attachments, and we are industrious creatures with needs for effectance, able to enter a state of vital engagement with our work. We are the rider and we are the elephant, and our mental health depends on the two working together, each drawing on the others' strengths. I don't believe there is an inspiring answer to the question, "What is the purpose *of* life?" Yet by drawing on ancient wisdom and modern science, we can find compelling answers to the question of purpose *within* life. The final version of the happiness hypothesis is that happiness comes from between. Happiness is not something that you can find, acquire, or achieve directly. You have to get the conditions right and then wait. Some of those conditions are within you, such as coherence among the parts and levels of your personality.

Other conditions require relationships to things beyond you: Just as plants need sun, water, and good soil to thrive, people need love, work, and a connection to something larger. It is worth striving to get the right relationships between yourself and others, between yourself and your work, and between yourself and something larger than yourself. If you get these relationships right, a sense of purpose and meaning will emerge.

# 11

## Conclusion: On Balance

*All things come into being by conflict of opposites.*

—HERACLITUS,[1] C. 500 BCE

*Without Contraries is no progression. Attraction and Repulsion, Reason and Energy, Love and Hate, are necessary to Human existence.*

—WILLIAM BLAKE,[2] C. 1790

THE ANCIENT CHINESE SYMBOL of yin and yang represents the value of the eternally shifting balance between seemingly opposed principles. As the epigrams above from Heraclitus and Blake show, this is not just an Eastern idea; it is Great Idea, a timeless insight that in a way summarizes the rest of this book. Religion and science, for example, are often thought to be opponents, but as I have shown, the insights of ancient religions and of modern science are both needed to reach a full understanding of human nature and the conditions of human satisfaction. The ancients may have known little about biology, chemistry, and physics, but many were good psychologists. Psychology and religion can benefit by taking each other seriously, or at least by agreeing to learn from each other while overlooking the areas of irreconcilable difference.

The Eastern and Western approaches to life are also said to be opposed: The East stresses acceptance and collectivism; the West encourages striving and individualism. But as we've seen, both perspectives are valuable. Happiness requires changing yourself and changing your world. It requires pursuing your own goals and fitting in with others. Different people at different times in their lives will benefit from drawing more heavily on one approach or the other.

And, finally, liberals and conservatives are opponents in the most literal sense, each using the myth of pure evil to demonize the other side and unite their own. But the most important lesson I have learned in my twenty years of research on morality is that nearly all people are morally motivated. Selfishness is a powerful force, particularly in the decisions of individuals, but whenever *groups* of people come together to make a sustained effort to change the world, you can bet that they are pursuing a vision of virtue, justice, or sacredness. Material self-interest does little to explain the passions of partisans on issues such as abortion, the environment, or the role of religion in public life. (Self-interest certainly cannot explain terrorism, but the selflessness made possible by group selection can.)

An important dictum of cultural psychology is that each culture develops expertise in some aspects of human existence, but no culture can be expert in all aspects. The same goes for the two ends of the political spectrum. My research[3] confirms the common perception that liberals are experts in thinking about issues of victimization, equality, autonomy, and the rights of individuals, particularly those of minorities and nonconformists. Conservatives, on the other hand, are experts in thinking about loyalty to the group, respect for authority and tradition, and sacredness.[4] When one side overwhelms the other, the results are likely to be ugly. A society without liberals would be harsh and oppressive to many individuals. A society without conservatives would lose many of the social structures and constraints that Durkheim showed are so valuable. Anomie would increase along with freedom. A good place to look for wisdom, therefore, is where you least expect to find it: in the minds of your opponents. You already know the ideas common on your own side. If you can take off the blinders of the myth of pure evil, you might see some good ideas for the first time.

By drawing on wisdom that is balanced—ancient and new, Eastern and Western, even liberal and conservative—we can choose directions in life that will lead to satisfaction, happiness, and a sense of meaning. We can't simply select a destination and then walk there directly—the rider does not have that much authority. But by drawing on humanity's greatest ideas and best science, we can train the elephant, know our possibilities as well as our limits, and live wisely.

# Acknowledgments

THIS BOOK EMERGED FROM my relationships with many people, which developed as I passed through four supportive universities. If this book is broader in its scope than most in psychology it is because I had the great fortune to be mentored by John Fisher at Yale, John Baron, Alan Fiske, Rick McCauley, Judith Rodin, Paul Rozin, and John Sabini at the University of Pennsylvania, and Richard Shweder at the University of Chicago. As an assistant professor at the University of Virginia, I received further mentoring from Dan Wegner, and also from Marty Seligman back at Penn. I am forever grateful to these generous teachers and broad-minded thinkers.

Books also require that somebody besides the author sees a possibility and takes a chance. I am deeply grateful to Sir John Templeton, the John Templeton Foundation, and its executive vice president, Arthur Schwartz, for supporting my research on moral elevation and for giving me a semester of sabbatical leave to begin the research for this book. My agent, Esmond Harmsworth, also took a chance; he invested a great deal of time and skill in guiding a first-time author through the complexities of the publishing world, and then to a partnership with editor Jo Ann Miller at Basic Books. Jo Ann encouraged me to write this book long before she became my editor, and she has improved the book in countless ways. Above all she helped me to aim high while writing accessibly, and I know my academic writings will benefit from her wisdom. I thank all these risk takers.

Many friends and colleagues read chapters and saved me from errors, overstatements, and puns. Jesse Graham, Suzanne King, Jayne Riew, and

Mark Shulman gave me detailed comments on the entire manuscript. The following people helped me improve one or more chapters: Jonathan Adler, Sara Algoe, Desiree Alvarez, Jen Bernhards, Robert Biswas-Diener, David Buss, Fredrik Bjorklund, Jerry Clore, William Damon, Judy Deloache, Nick Epley, Sterling Haidt, Greg LaBlanc, Angel Lillard, Bill McAllister, Rick McCauley, Helen Miller, Brian Nosek, Shige Oishi, James Pawelski, Paul Rozin, Simone Schnall, Barry Schwartz, Patrick Seder, Gary Sherman, Nina Strohminger, Bethany Teachman, Kees Van den Bos, Dan Wegner, Dan Willingham, Nancy Weinfield, Emily Wilson, and Tim Wilson. I thank them all.

Finally, a book emerges from the personality of its author, and whether personality is shaped by nature or nurture, I thank my parents, Harold and Elaine Haidt, as well as my sisters, Rebecca Haidt and Samantha Davenport, for their loving support. Above all I thank my wife, Jayne Riew, who gave me a between.

# Notes

## INTRODUCTION: TOO MUCH WISDOM

1. From *Hamlet,* II.ii.249–250. All quotations from Shakespeare are from G. Blakemore (Ed), 1974. *The Riverside Shakespeare* (Boston: Houghton Mifflin).

2. Seligman, 2002.

3. Keyes and Haidt, 2003.

4. Technically one should say "The Buddha" (the awakened one), just as one should say "The Christ" (the anointed one). However, I will follow common usage in referring to Buddha and Christ.

## CHAPTER 1

1. This and all subsequent quotations from the Old and New Testaments are from the New Revised Standard Version.

2. Franklin, 1980/1733–1758, 3.

3. Lakoff and Johnson, 1980.

4. *Dhammapada,* verse 326, in Mascaro, 1973.

5. Plato, *Phaedrus* 253d, in Cooper, 1997.

6. Freud, 1976/1900.

7. Ovid, *Metamorphoses*, Bk. VII, 249.

8. Montaigne, 1991/1588, 115. The second quote is also from page 115.

9. Gershon, 1998.

10. Lyte, Varcoe, and Bailey, 1998.

11. Gazzaniga, 1985; Gazzaniga, Bogen, and Sperry, 1962.

12. Gazzaniga, 1985, 72.

13. Feinberg, 2001.

14. Olds and Milner, 1954.

15. Burns and Swerdlow, 2003.

16. Damasio, 1994; Rolls, 1999.

17. Rolls, 1999.

18. For summaries of findings on the "emotional brain" see Berridge, 2003; LeDoux, 1996.

19. Damasio, 1994, Damasio, Tranel, and Damasio, 1990.

20. Bargh, Chen, and Burrows, 1996.

21. Bargh et al., 1996, for the elderly effect; Dijksterhuis and van Knippenberg, 1998, for the others.

22. James, 1950/1890.

23. See review in Leakey, 1994.

24. For a review of why most mental systems work so well, yet logical reasoning works so poorly, see Margolis, 1987.

25. Rolls, 1999.

26. Hume, 1969/1739, 462.

27. Shoda, Mischel, and Peake, 1990.

28. For a review of these studies and a full account of the interplay between the hot (automatic) and cool (controlled) systems, see Metcalfe and Mischel, 1999.

29. Salovey and Mayer, 1990. Possessing emotional intelligence does *not* mean that one's emotions are intelligent.

30. Baumeister et al., 1998.

31. Obeyesekere, 1985.

32. Wegner, 1994.

33. Haidt, 2001; Haidt, Koller, and Dias, 1993.

34. Gladwell, 2005.

CHAPTER 2

1. *Meditations*, 4:3.

2. *Dhammapada*, verse 1, in Mascaro, 1973.

3. Carnegie, 1984/1944, 113.

4. From Dr. Phil's "Ten Life Laws," retrieved from www.drphil.com on 12/16/04.

5. Boethius, 1962/c. 522 CE, 24.

6. Boethius, 1962/c. 522 CE, 22.

7. Boethius, 1962/c. 522 CE, 29.

8. See Miller and C'de Baca, 2001, for a review.

9. Bargh et al., 1996; Fazio et al., 1986.

10. Nosek, Banaji, and Greenwald, 2002; Nosek, Greenwald, and Banaji, in press.

11. Pelham, Mirenberg, and Jones, 2002.

12. Pinker, 1997.

13. See two recent reviews: Baumeisteret et al., 2001; Rozin and Royzman, 2001.

14. Gottman, 1994.

15. Kahneman and Tversky, 1979.

16. Rozin and Royzman, 2001.

17. Franklin, 1980/1733–1758, 26.

18. Gray, 1994; Ito and Cacioppo, 1999.

19. Miller, 1944.

20. LaBar and LeDoux, 2003.

21. Shakespeare, *Hamlet*, I.ii.133–134.

22. Shakespeare, *Hamlet*, II.ii.249–250.

23. Angle and Neimark, 1997.

24. Lykken et al., 1992.

25. Bouchard, 2004; Plomin and Daniels, 1987; Turkheimer, 2000.

26. Marcus, 2004.

27. Plomin and Daniels, 1987.

28. Lykken and Tellegen, 1996.

29. Davidson, 1998.

30. Davidson and Fox, 1989.

31. Kagan, 1994; Kagan, 2003.

32. Milton, *Paradise Lost* bk. 1, lines 254–255.

33. See Shapiro, Schwartz, and Santerre, 2002, for a review. Most of the published studies on meditation have used weak or flawed designs (such as comparing people who chose to sign up for a meditation class with people who did not). However, Shapiro et al. review several studies that used random assignment to either a meditation condition or a control condition. The benefits I mention in the text are those supported by studies that used random assignment.

34. Definition from Shapiro et al., 2002.

35. *Dhammapada*, verse 205, in Mascaro, 1973.

36. Beck, 1976.

37. Dobson, 1989; Hollon and Beck, 1994.

38. DeRubeis et al., 2005.

39. Seligman, 1995.

40. An easy place to start is with the popular book *Feeling Good* by David Burns, 1999. Just reading this book has been shown to be an effective treatment for depression (Smith et al., 1997).

41. Proust, 1992/1922b, 291.

42. Nestler, Hyman, and Malenka, 2001.

43. Schatzberg, Cole, and DeBattista, 2003. Occasional reports that SSRIs are no more effective than placebos appear to be based on flawed studies; for example, studies that used very low doses of SSRIs. See Hollon et al., 2002.

44. Kramer, 1993.

45. Haidt, 2001; Haidt and Joseph, 2004.

# CHAPTER 3

1. *Analects*, 15.24. In Leys, 1997.

2. Babylonian Talmud, Tractate Shabbos, Folio 31a, Schottenstein edition, A. Dicker, trans. (New York: Mesorah Publications, 1996).

3. *The Godfather*, directed by F. F. Coppola, 1972. Paramount Pictures. Based on the novel by Mario Puzo.

4. Campbell, 1983; Richerson and Boyd, 1998.

5. Hamilton, 1964, first worked out the details of kin selection. We all share most of our genes with all people, and even with most chimpanzees, mice, and fruit flies. What matters here is only the subset of genes that vary within the human population.

6. Of course, the ancestors did no "parlaying"; they just survived better than their competitors, and in the process, reproduction shifted over to a queen and ultrasociality emerged.

7. Described in Ridley, 1996.

8. Kunz and Woolcott, 1976.

9. Cialdini, 2001.

10. Axelrod, 1984.

11. Wilkinson, 1984.

12. Trivers, 1971.

13. Ridley, 1996.

14. Panthanathan and Boyd, 2004; Richerson and Boyd, 2005.

15. Cosmides and Tooby, 2004.

16. Guth, Schmittberger, and Schwarze, 1982.

17. Sanfey et al., 2003.

18. Bjorklund, 1997.

19. Dunbar, 1993.

20. Dunbar, 1996.

21. Hom and Haidt, in preparation.

22. For a defense of gossip, see Sabini and Silver, 1982.

23. Cialdini, 2001.

24. Cialdini, 2001, cites an unpublished study by Lynn and McCall, 1998.

25. James and Bolstein, 1992.

26. Cialdini et al., 1975.

27. Benton, Kelley, and Liebling, 1972.

28. Lakin and Chartrand, 2003.

29. van Baaren et al., 2004.

30. van Baaren et al., 2003.

## CHAPTER 4

1. *Dhammapada*, verse 252, in Mascaro, 1973.

2. "Outing Mr. Schrock," *Washington Post*, September 2, 2004, A22.

3. Hom and Haidt, in preparation.

4. For extensive discussions of the prisoner's dilemma game, see Axelrod, 1984; Wright, 1994.

5. Machiavelli, *The Discourses*, 1.25.

6. Byrne and Whiten, 1988.

7. Batson et al., 1997; Batson et al., 1999.

8. Buchanan, 1965, 53.

9. Pachocinski, 1996, 222.

10. Wright, 1994, 13.

11. Kuhn, 1991.

12. Perkins, Farady, and Bushey, 1991.

13. Kunda, 1990; Pyszczynski and Greenberg, 1987.

14. Franklin, 1962/c. 1791, 43.

15. Alicke et al., 1995; Hoorens, 1993.

16. Heine and Lehman, 1999; Markus and Kitayama, 1991.

17. Epley and Dunning, 2000.

18. This analysis of leadership, and the studies cited in this paragraph come from Dunning, Meyerowitz, and Holzberg, 2002.

19. Cross, 1977.

20. Taylor et al., 2003.

21. Ross and Sicoly, 1979.

22. Epley and Caruso, 2004.

23. Babcock and Loewenstein, 1997.

24. Pronin, Lin, and Ross, 2002.

25. Hick, 1967.

26. Russell, 1988; Boyer, 2001.

27. Baumeister, 1997.

28. See review in Baumeister, 1997 (chap. 2).

29. Baumeister, Smart, and Boden, 1996; Bushman and Baumeister, 1998. However, evidence that antisocial behavior is associated with *low* self-esteem has recently been reported by Donnellan et al., 2005.

30. Glover, 2000.

31. Skitka, 2002.

32. Geertz, 1973, 5, paraphrasing the sociologist Max Weber.

33. *Bhagavad Gita*, 12.18–19. In Zaehner, 1969.

34. Sent-ts'an, *Hsin hsin ming*. In Conze, 1954.

35. Shapiro et al., 2002.

36. Burns, 1999.

## CHAPTER 5

1. *Dhammapada*, verse 83, in Mascaro, 1973.

2. Epictetus, 1983/1st–2nd cent. CE, 9.

3. Davidson, 1994; see also Brim, 1992.

4. *Troilus and Cressida*, I.ii.287.

5. Wilson and Gilbert, 2003.

6. Brickman, Coates, and Janoff-Bulman, 1978; see also Schulz and Decker, 1985, for long-term follow-up of spinal injury patients. No study has obtained happiness or life satisfaction ratings in the first days after winning the lottery or becoming a paraplegic, but appearances suggest that emotional reactions are very strong. We can therefore infer that the surprisingly moderate happiness ratings given by both groups a few months later illustrate a return "most of the way" to baseline.

7. Kaplan, 1978.

8. Interview by Deborah Solomon, *New York Times Magazine,* Sunday December 12, 2004, 37. It should be noted, however, that adaptation to severe disability is slow and often incomplete. Even years later, paraplegics have not, on average, returned fully to their pre-accident levels.

9. Helson, 1964.

10. For a sensitive exploration of goal pursuit, ambition, and happiness, see Brim, 1992.

11. Lykken and Tellegen, 1996.

12. Smith, 1976/1759, 149.

13. Brickman and Campbell, 1971.

14. Diener et al., 1999; Mastekaasa, 1994; Waite and Gallagher, 2000. However, it is not clear that married people are, on average, happier than those who never married, because unhappily married people are the least happy group of all and they pull down the average; see DePaulo and Morris, 2005, for a critique of research on the benefits of marriage.

15. Harker and Keltner, 2001; Lyubomirsky, King, and Diener, in press.

16. Baumeister and Leary, 1995. However, it is not certain that marriage itself is more beneficial than other kinds of companionship. Much evidence says yes, particularly for health, wealth, and longevity (reviewed in Waite and Gallagher, 2000); but a large longitudinal study failed to find a long-lasting benefit of marriage on reports of well-being (Lucas et al., 2003).

17. Diener et al., 1999; Myers, 2000.

18. Argyle, 1999. Some studies find a larger race difference, but when differences in income and job status are controlled for, the differences become small or insignificant.

19. Diener et al., 1999; Lucas and Gohm, 2000.

20. Carstensen et al., 2000; Diener and Suh, 1998. Mroczek and Spiro, 2005, found a peak around age sixty-five.

21. Frederick and Loewenstein, 1999; Riis et al., 2005.

22. Lucas, 2005.

23. Schkade and Kahneman, 1998.

24. Feingold, 1992.

25. Diener, Wolsic, and Fujita, 1995.

26. Diener and Oishi, 2000.

27. Lyubomirsky, King, and Diener, in press; Fredrickson, 2001.

28. Diener and Oishi, 2000; Frank, 1999.

29. *Bhagavad Gita*, XVI.12. The second quote is from XVI.13–14. In Zaehner, 1969.

30. Plomin and Daniels, 1987. The *unique* environment that each child creates within the family matters, but not usually as much as his or her unique genes.

31. Lykken, 1999.

32. Marcus, 2004.

33. Lyubomirsky, Sheldon, and Schkade, in press.

34. See Lyubomirsky et al., in press, and Seligman, 2002, chap. 4. Lyubomirsky et al. call the last term "activities"; Seligman calls it "voluntary variables." I am combining their terms, for simplicity of explanation, by referring to "voluntary activities."

35. Glass and Singer, 1972, and others reviewed in Frederick and Loewenstein, 1999.

36. See review in Frank, 1999.

37. Koslowsky and Kluger, 1995.

38. Csikszentmihalyi, 1997.

39. Glass and Singer, 1972.

40. Langer and Rodin, 1976; Rodin and Langer, 1977.

41. Haidt and Rodin, 1999.

42. Reviewed in Lyubomirsky, King, and Diener, in press; Reis and Gable, 2003.

43. See Argyle, 1999; Baumeister and Leary, 1995; Myers, 2000; Seligman, 2002. However, Lucas and Dyrenforth (in press) present evidence that the direct causal effect of improved social relationships on happiness may be smaller than most psychologists realize, perhaps no larger than the effect of income on happiness. This debate has just begun; its resolution must await future research.

44. Lyubomirsky, King, and Diener, in press; Reis and Gable, 2003.

45. Frederick and Loewenstein, 1999.

46. Bronte, 1973/1847, 110. Spoken by Jane Eyre.

47. Belk, 1985; Kasser, 2002; Kasser and Ryan, 1996.

48. Csikszentmihalyi, 1990.

49. See Miller, 1997, on the "disgust of surfeit."

50. Seligman, 2002, 102.

51. Wrzesniewski, Rozin, and Bennett, 2003; see also Kass, 1994.

52. Epicurus, *Letter to Menoeceus,* 126. In O'Connor, 1993.

53. Peterson and Seligman, 2004.

54. Emmons and McCullough, 2003; Lyubomirsky, Sheldon, and Schkade, in press.

55. Frank, 1999.

56. Adapted from Solnick and Memenway, 1998.

57. Van Boven and Gilovich, 2003.

58. *Tao Te Ching,* 12, in Feng and English, 1972.

59. This same argument has been made with neuroscientific evidence by Whybrow, 2005.

60. Iyengar and Lepper, 2000.

61. Schwartz, 2004.

62. Schwartz et al., 2002.

63. Schwartz et al., 2002.

64. Conze, 1959.

65. Conze, 1959, 40.

66. Some people say "the Buddha" (the awakened one), just as some people say "the Christ" (the anointed one). However I follow common usage in referring to Buddha and Christ.

67. Biswas-Diener and Diener, 2001; Diener and Diener, 1996.

68. Biswas-Diener and Diener, 2001, 337.

69. I later found a published version of the talk: Solomon, 1999.

70. Broderick, 1990, 261.

71. Memorial Day Address, delivered on May 30, 1884. In Holmes, 1891, 3.

## CHAPTER 6

1. Seneca, Epistle XLVIII, in Seneca, 1917–1925/c. 50 CE, 315.

2. Meditation XVII, in Donne, 1975/1623.

3. The facts in this paragraph are drawn from Blum, 2002, Chapter 2.

4. Watson, 1928.

5. My account of Harlow's career is taken from Blum, 2002.

6. Harlow, Harlow, and Meyer, 1950.

7. Harlow and Zimmerman, 1959.

8. Blum, 2002.

9. For reviews of the development of Bowlby's life and ideas, see Blum, 2002, and Cassidy, 1999.

10. Lorenz, 1935.

11. Bowlby, 1969; Cassidy, 1999.

12. For a review of the functions of play, see Fredrickson, 1998.

13. Harlow, 1971.

14. Ainsworth et al., 1978.

15. See current reviews of attachment research in Cassidy, 1999; Weinfield et al., 1999.

16. Harris, 1995.

17. Kagan, 1994.

18. DeWolff and van IJzendoorn, 1997.

19. van IJzendoorn et al., 2000.

20. Hazan and Shaver, 1987. Copyright © 1987 by the American Psychological Association. Adapted with permission.

21. Hazan and Zeifman, 1999.

22. Feeney and Noller, 1996.

23. Bowlby, 1969.

24. Hazan and Zeifman, 1999.

25. Vormbrock, 1993.

26. Carter, 1998; Uvnas-Moberg, 1998.

27. Taylor et al., 2000.

28. See Fisher, 2004, for a review of oxytocin's role in love and sex.

29. Fisher, 2004.

30. Moss, 1998.

31. Trevathan, 1987; Bjorklund, 1997.

32. Bjorklund, 1997.

33. Hill and Hurtado, 1996.

34. Buss, 2004.

35. Jankowiak and Fischer, 1992.

36. Berscheid and Walster, 1978; see also Sternberg, 1986.

37. Plato, *Symposium 192e,* A. Nehamas and P. Woodruff (trans.). In Cooper, 1997.

38. Berscheid and Walster, 1978.

39. Quoted by Jankowiak and Fischer, 1992.

40. Julien, 1998.

41. Bartels and Zeki, 2000; Fisher, 2004.

42. These are the three components of Sternberg's (1986) triangular theory of love.

43. *Dhammapada,* verse 284, in Mascaro, 1973.

44. Chap. 2, line 213, in Doniger and Smith, 1991.

45. *Analects* 9.18, in Leys, 1997.

46. Tantric traditions may seem to be ancient exceptions, but their goal was to use the energy of lust and other passions, often in conjunction with disgust, as a way to break attachments to carnal pleasures. See Dharmakirti, 2002.

47. Plato, *Symposium 192e,* A. Nehamas and P. Woodruff (trans.). In Cooper, 1997.

48. Plato, *Symposium 210d,* A. Nehamas and P. Woodruff (trans.). In Cooper, 1997.

49. Lucretius, *De Rerum Natura,* bk. IV, lines 1105–1113.

50. Goldenberg et al., 2001; Goldenberg et al., 1999.

51. Becker, 1973; Pyszcsynski, Greenberg, and Solomon, 1997.

52. Durkheim, 1951/1897, 209.

53. See reviews in Cohen and Herbert, 1996, Waite and Gallagher, 2000. However, Lucas and Dyrenforth (in press) have recently questioned whether social relationships are quite as important as the rest of the field thinks.

54. Fleeson, Malanos, and Achille, 2002.

55. Brown et al., 2003.

56. Baumeister and Leary, 1995.

57. Sartre, 1989/1944, 45.

## CHAPTER 7

1. Known also as Mencius. From *The Book of Mencius*, section 6B:15, in Chan, 1963, 78.

2. Nietzsche, 1997/1889, 6.

3. Taylor, 2003.

4. This story is true, but names and identifying details have been changed.

5. Cleckley, 1955; Hare, 1993.

6. For reviews of posttraumatic growth see Nolen-Hoeksema and Davis, 2002; Tedeschi, Park, and Calhoun, 1998; Tennen and Affleck, 1998; Updegraff and Taylor, 2000. There were a few early pioneers, such as Frankl, 1984/1959.

7. Meichenbaum, 1985, reviewed in Updegraff and Taylor, 2000.

8. Dalai Lama, 2001/1995, 40.

9. Nolen-Hoeksema and Davis, 2002, 602–603.

10. Baum, 2004; Tennen and Affleck, 1998.

11. *As You Like It*, II.i.12–14.

12. Tooby and Cosmides, 1996.

13. Costa and McCrae, 1989.

14. Park, Cohen, and Murch, 1996.

15. Costa and McCrae, 1989.

16. Srivastava et al., 2003.

17. McAdams, 1994; McAdams, 2001.

18. McAdams, 1994, 306.

19. Emmons, 2003; Emmons, 1999.

20. See also the work Tim Kasser: Kasser, 2002; Kasser and Ryan, 1996.

21. McAdams, 2001, 103.

22. Adler, Kissel, and McAdams, in press.

23. Sheldon and Kasser, 1995.

24. See Emmons, 2003, chap. 6; and James, 1961/1902.

25. See King, 2001, on the "hard road to the good life."

26. Lerner and Miller, 1978.

27. For new research on sense making as part of the "psychological immune system" see Wilson and Gilbert, 2005.

28. Nolen-Hoeksema and Davis, 2002; Ryff and Singer, 2003; Tennen and Affleck, 1998. Other traits that matter, though less than optimism, are cognitive complexity and openness to experience.

29. Carver, Scheier, and Weintraub, 1989; Lazarus and Folkman, 1984.

30. Pennebaker, 1997.

31. Tavris, 1982.

32. Pennebaker, 1997, 99–100.

33. Myers, 2000; McCullough et al., 2000.

34. Pennebaker, 1997.

35. Chorpita and Barlow, 1998.

36. See Belsky, Steinberg, and Draper, 1991, for a variety of psychological and biological changes wrought by early stressful environments.

37. Rind, Tromovitch, and Bauserman, 1998.

38. McAdams, 2001.

39. Fitzgerald, 1988.

40. Elder, 1974; Elder, 1998.

41. I interviewed Elder in 1994 for a report for the MacArthur Foundation.

42. Durkheim, 1951/1897.

43. Putnam, 2000.

44. Baltes, Lindenberger, and Staudinger, 1998.

45. Proust, 1992a/1922, 513.

46. Sternberg, 1998; see also Baltes and Freund, 2003.

47. The theologian Reinhold Niebuhr used a variant of this prayer in a sermon in 1943, and this is thought by some to be the source of the version given here, which was popularized by Alcoholics Anonymous.

## CHAPTER 8

1. Epicurus, *Principle Doctrines*. In Epicurus, 1963/c. 290 BCE, 297.

2. *Dhammapada*, sec. 9, stanza 118. This translation is from Byrom, 1993. It has the same meaning as the translation in Mascaro, but has much better flow.

3. Aristotle, 1962/4th cent. BCE, 1098a.

4. Franklin, 1962/c. 1791, 82.

5. Franklin, 1962/c. 1791, 82.

6. Franklin, 1962/c. 1791, 88.

7. Peterson and Seligman, 2004.

8. In Lichtheim, 1976, 152.

9. Templeton, 1997.

10. Hansen, 1991.

11. Aristotle, 1962/4th cent. BCE, 1103b.

12. Kant, 1959/1785.

13. Bentham, 1996/1789.

14. Pincoffs, 1986.

15. M. B. Sure, "Raising a Thinking Child Workbook," retrieved on April 15, 2005, from www.thinkingchild.com.

16. Singer, 1979.

17. MacIntyre, 1981.

18. See also Taylor, 1989.

19. Peterson and Seligman, 2004.

20. Piaget, 1965/1932.

21. Shweder et al., 1997.

22. Baumeister, 1997, discussed in chapter 4.

23. Webster's *New Collegiate Dictionary*, 1976.

24. Lyubomirsky et al., in press.

25. Isen and Levin, 1972. There are limits on this effect, such as when the helping will ruin the happy mood, Isen and Simmonds, 1978.

26. Piliavin, 2003.

27. Thoits and Hewitt, 2001.

28. Brown et al., 2003.

29. McAdams, 2001, discussed in chapter 7.

30. Piliavin, 2003.

31. Emmons, 2003.

32. Durkheim, 1951/1897, discussed in chapter 6.

33. Sampson, 1993.

34. Hunter, 2000.

35. Appiah, 2005. See also Taylor, 1989.

36. Tajfel, 1982.

37. Haidt, Rosenberg, and Hom, 2003.

38. Damon, 1997.

## CHAPTER 9

1. Formerly known as Mencius. Quoted in Chan, 1963, 59.

2. From the Hadith, quoted in Fadiman and Frager, 1997, 6.

3. Abbott, 1952/1884. The extended quote is from page 80.

4. Boehm, 1999.

5. Brown and Gilman, 1960.

6. See Leviticus 12; Buckley and Gottlieb, 1988.

7. Rozin and Fallon, 1987.

8. Rozin et al., 1997.

9. Leakey, 1994.

10. For a review of our research on disgust, see Rozin, Haidt, and McCauley, 2000.

11. Haidt et al., 1997.

12. Reported in Thomas, 1983, 38.

13. John Wesley, 1984/1786, sermon 88, "On Dress," 249.

14. Shweder et al., 1997.

15. Haidt, Koller, and Dias, 1993.

16. Doniger and Smith, 1991. The long quote is from chap. 4, stanzas 109–122.

17. See Bloom, 2004, on how people are "natural born dualists," keeping body and soul apart.

18. From "The Divinity School Address," in Emerson, 1960/1838, 102.

19. Stall, 1897. The quote is from page 35 of the 1904 edition.

20. Steele, 1867, 191.

21. Le Conte, 1892, 330.

22. Eliade, 1959/1957. The long quote is from page 24.

23. Based on the seminal work of Ekman, Sorensen, and Friesen, 1969.

24. Jefferson, 1975/1771.

25. Isen and Levin, 1972; see discussion in chap. 8.

26. Algoe and Haidt, 2005.

27. Thrash and Elliot, 2004.

28. McCraty and Childre, 2004.

29. Carter, 1998, and see chap. 6.

30. See a recent finding that oxytocin increases trust, Kosfeld, et al., 2005.

31. David Whitford, personal communication, 1999. Used with permission.

32. See discussion of attachment and agape in chap. 6.

33. From the *Critque of Practical Reason*, quoted in Guyer, 1992, 1.

34. From Darwin's "Autobiography," quoted in Wright, 1994, 364.

35. From *Nature*, in Emerson, 1960b/1838, 24.

36. Wasson, 1986.

37. Shulgin and Shulgin, 1991.

38. Grob and de Rios, 1994.

39. Pahnke, 1966.

40. Keltner and Haidt, 2003.

41. *Bhagavad Gita*, 2.45. In Zaehner, 1969.

42. James, 1961/1902.

43. James, 1961/1902, 216–217.

44. Maslow, 1964.

45. Daston and Park, 1998.

46. Maslow, 1964, 58.

47. Leary, 2004.

48. Gallup, 1982.

49. Quoted in Cruikshank, 1999, 95.

50. Warren, 2002.

51. I have extended Shweder's three ethics into a theory of five foundations of intuitive ethics, which I use to analyze the culture war. See Haidt and Bjork-lund, in press; Haidt and Joseph, 2004.

52. Warren, 2002, 22.

53. Haidt and Hersh, 2001.

54. Gross and Haidt, 2005.

55. Haidt and Hersh, 2001, 208.

## CHAPTER 10

1. Isa Upanishad, verses 6–7. In Mascaro, 1965, 49–50.

2. Spoken by Jim in *My Antonia;* Cather, 1987/1918, 14.

3. "On the Road to Find Out" by Cat Stevens. From the album "Tea for the Tillerman," 1970, A&M.

4. Letter to John Augustine Washington, in Irving, 1976/1856–1859.

5. "Sherry Darling" by Bruce Springsteen. Copyright © 1980 Bruce Springsteen (ASCAP). Reprinted by permission. International copyright secured. All rights reserved.

6. Allen, 1975.

7. See Klemke, 2000, for a volume of philosophical essays on the meaning of life. Most of the nontheistic essays try to do just this.

8. For examples, see Appiah, 2005; Churchland, 1998; Flanagan, 1991; Gibbard, 1990; Nussbaum, 2001; Solomon, 1999.

9. Adams, 1980.

10. *Monty Python's The Meaning of Life,* directed by Terry Gilliam (Universal Studios, 1983).

11. *Webster's Third New International Dictionary,* 1993, unabridged, for both words.

12. Jung, 1963.

13. *Nichomachean Ethics*, bk. 1, 1094a.

14. Warren, 2002.

15. Bonanno, 2004, and see chap. 7.

16. Gardner, Csikszentmihalyi, and Damon, 2001.

17. A well-respected theory, Ryan and Deci, 2000, says that the fundamental psychological needs are competence (including work), relatedness (love), and autonomy. I agree that autonomy is important, but I don't think it is as important, universal, or consistently good as the other two.

18. This phrase, *"lieben und arbeiten,"* does not appear in Freud's writings. It is often claimed to be something Freud once said in a conversation. Erik Erikson reports it in this way in Erikson, 1963/1950, 265.

19. Leo Tolstoy, quoted in Troyat, 1967, 158.

20. White, 1959.

21. White, 1959, 322.

22. *Troilus and Cressida*, I.ii.287.

23. Marx, 1977/1867.

24. Kohn and Schooler, 1983.

25. Bellah et al., 1985.

26. Wrzesniewski et al., 2003; Wrzesniewski, Rozin, and Bennett, 2003.

27. As discussed in chap. 8.

28. Fredrickson, 2001.

29. Gibran, 1977/1923, 27.

30. Nakamura and Csikszentmihalyi, 2003, 87.

31. Nakamura and Csikszentmihalyi, 2003, 86.

32. Gardner, Csikszentmihalyi, and Damon, 2001. See also Damon, Menon, and Bronk, 2003, on the development of purpose.

33. For example, Fenton, 2005.

34. Much recent work in psychology shows the importance of fit or coherence for well-being. See Freitas and Higgins, 2002; Tamir, Robinson, and Clore, 2002.

35. Emmons, 1999; Miller and C'de Baca, 2001.

36. For a well-developed multilevel approach to "optimal human being," see Sheldon, 2004.

37. I'm drawing here from interdisciplinary work in cognitive science on the role of the body and of culture in cognition, such as that of Clark, 1999; Lakoff and Johnson, 1999; and Shore, 1996.

38. Durkheim, 1965/1915; Wilson, 2002.

39. Brown, 1991.

40. Darwin, 1998/1871, 166.

41. Williams, 1966; Trivers, 1971.

42. Dawkins, 1976.

43. Wilson, 1990.

44. Camponotus saundersi, described in Wilson, 1990, 44.

45. Wilson, 2002. But note that group selection is quite controversial, and it is presently a minority position among evolutionary biologists.

46. See Aunger, 2000; Gladwell, 2000; Richerson and Boyd, 2005.

47. Richerson and Boyd, 2005; Leakey, 1994.

48. Mithen, 2000, explains the gap between the brain's reaching its current size, over 100,000 years ago, and the cultural explosion that began a few tens of thousands of years later as a result of slowly accumulating material culture.

49. See Pinker, 1997, 2002, on how the evolved mind constrains the arts, politics, gender roles, and other aspects of culture.

50. Foxes have been domesticated and made somewhat dog-like in appearance and behavior in just forty years of selective breeding; see Belyaev, 1979; Trut, 1999.

51. Richerson and Boyd, 2005.

52. Durkheim, 1965/1915, 62.

53. Boyer, 2001.

54. Boyer, 2001; Dawkins, 1976.

55. Hamer, 2004.

56. The term had recently been coined by R. M. Bucke. See James, 1961/1902, 313.

57. From the *Columbia Encyclopedia*, 6th edition, 2001. Entry for "yoga."

58. Quoted by James, 1961/1902, 317.

59. Newberg, D'Aquili, and Rause, 2001.

60. McNeill, 1995, 2.

61. McNeill, 1995.

62. From Gray, 1970/1959, quoted on p. 10 of McNeill, 1995.

## CHAPTER 11

1. Quoted by Diogenes Laertius, 1925/3rd cent. CE, bk. 9, sec. 8

2. Blake, 1975/1790–1793, 3.

3. Graham and Haidt, in preparation; Haidt and Bjorklund, in press; Haidt and Hersh, 2001.

4. There are, of course, subtypes of liberals and conservatives that violate these generalizations, such as the religious left and the libertarian right, each with its own expertise.

# References

Abbott, E. A. (1952/1884). *Flatland: A romance of many dimensions*. (6th ed.). New York: Dover.

Adams, D. (1980). *The hitchhiker's guide to the galaxy*. New York: Harmony Books.

Adler, J. M., Kissel, E., and McAdams, D. P. (in press). Emerging from the CAVE: Attributional style and the narrative study of identity in midlife adults. *Cognitive Therapy and Research*.

Ainsworth, M. D. S., Blehar, M., Waters, E. & Wall, S. (1978). *Patterns of attachment: A psychological study of the strange situation*. Hillsdale, NJ: Erlbaum.

Algoe, S., and Haidt, J. (2005). Witnessing excellence in action: The "other-praising" emotions of elevation, gratitude, and admiration. Unpublished manuscript, University of Virginia.

Alicke, M. D., Klotz, M. L., Breitenbecher, D. L., Yurak, T. J., & Vredenburg, D. S. (1995). Personal contact, individuation, and the better-than-average effect. *Journal of Personality and Social Psychology, 68*, 804–825.

Allen, W. (1975). *Without feathers*. New York: Random House.

Angle, R., & Neimark, J. (1997). Nature's clone. *Psychology Today*, July/August.

Appiah, K. A. (2005). *The ethics of identity*. Princeton: Princeton University Press.

Argyle, M. (1999). Causes and correlates of happiness. In D. Kahneman, E. Diener & N. Schwartz (Eds.), *Well-being: The foundations of hedonic psychology* (pp. 353–373). New York: Russell Sage.

Aristotle. (1962/4th cent. BCE). *Nichomachean ethics* (M. Oswald, Trans.). Indianapolis, IN: Bobbs-Merrill.

Aunger, R. (Ed.). (2000). *Darwinizing culture: The status of memetics as a science*. Oxford, UK: Oxford University Press.

Aurelius, M. (1964/2nd cent. CE). *Meditations* (M. Staniforth, Trans.) London: Penguin.

Axelrod, R. (1984). *The evolution of cooperation*. New York: Basic Books.

Babcock, L., & Loewenstein, G. (1997). Explaining bargaining impasse: The role of self-serving biases. *Journal of Economic Perspectives*, 11, 109–126.

Baltes, P. B., & Freund, A. M. (2003). The intermarriage of wisdom and selective optimization with compensation: Two meta-heuristics guiding the conduct of life. In C. L. M. Keyes & J. Haidt (Eds.), *Flourishing: Positive psychology and the life well-lived* (pp. 249–273). Washington, DC: American Psychological Association.

Baltes, P. B., Lindenberger, U., & Staudinger, U. M. (1998). Life-span theory in developmental psychology. In W. Damon & R. Lerner (Eds.), *Handbook of child psychology. Vol. 1, Theoretical models of human development*. (5th ed.). (pp. 1029–1143). New York: Wiley.

Bargh, J. A., Chaiken, S., Raymond, P., & Hymes, C. (1996). The automatic evaluation effect: Unconditionally automatic activation with a pronunciation task. *Journal of Experimental Social Psychology*, 32, 185–210.

Bargh, J. A., Chen, M., & Burrows, L. (1996). Automaticity of social behavior: Direct effects of trait construct and stereotype activation on action. *Journal of Personality and Social Psychology*, 71, 230–244.

Bartels, A., & Zeki, S. (2000). The neural basis of romantic love. *Neuroreport*, 11, 3829–3834.

Batson, C. D., Kobrynowicz, D., Dinnerstein, J. L., Kampf, H. C., & Wilson, A. D. (1997). In a very different voice: Unmasking moral hypocrisy. *Journal of Personality and Social Psychology*, 72, 1335–1348.

Batson, C. D., Thompson, E. R., Seuferling, G., Whitney, H., & Strongman, J. A. (1999). Moral hypocrisy: Appearing moral to oneself without being so. *Journal of Personality and Social Psychology*, 77, 525–537.

Baum, D. (2004). The price of valor. *The New Yorker*, July 12.

Baumeister, R. F. (1997). *Evil: Inside human cruelty and violence*. New York: W. H. Freeman.

Baumeister, R. F., Bratlavsky, E., Finenauer, C., & Vohs, K. D. (2001). Bad is stronger than good. *Review of General Psychology*, 5, 323–370.

Baumeister, R. F., Bratlavsky, E., Muraven, M., & Tice, D. M. (1998). Ego depletion: Is the active self a limited resource? *Journal of Personality and Social Psychology*, 74, 1252–1265.

Baumeister, R. F., & Leary, M. R. (1995). The need to belong: Desire for interpersonal attachments as a fundamental human motivation. *Psychological Bulletin*, 117, 497–529.

Baumeister, R. F., Smart, L., & Boden, J. M. (1996). Relation of threatened egotism to violence and aggression: The dark side of high self-esteem. *Psychological Review*, 103, 5–33.

Beck, A. T. (1976). *Cognitive therapy and the emotional disorders*. New York: International Universities Press.

Becker, E. (1973). *The Denial of Death*. New York: Free Press.

Belk, R. W. (1985). Materialism: Trait aspects of living in the material world. *Journal of Consumer Research*, 12, 265–280.

Bellah, R., Madsen, R., Sullivan, W. M., Swidler, A., & Tipton, S. (1985). *Habits of the heart*. New York: Harper and Row.

Belsky, J., Steinberg, L., & Draper, P. (1991). Childhood experience, interpersonal development, and reproductive strategy: An evolutionary theory of socialization. *Child Development*, 62, 647–670.

Belyaev, D. K. (1979). Destabilizing selection as a factor in domestication. *Journal of Heredity*, 70, 301–308.

Bentham, J. (1996/1789). *An introduction to the principles of morals and legislation*. Oxford: Clarendon.

Benton, A. A., Kelley, H. H., & Liebling, B. (1972). Effects of extremity of offers and concession rate on the outcomes of bargaining. *Journal of Personality and Social Psychology*, 24, 73–83.

Berridge, K. C. (2003). Comparing the emotional brains of humans and other animals. In R. J. Davidson, K. R. Scherer & H. H. Goldsmith (Eds.), *Handbook of affective sciences* (pp. 25–51). Oxford, UK: Oxford University Press.

Berscheid, E., & Walster, E. H. (1978). *Interpersonal attraction*. New York: Freeman.

Biswas-Diener, R., & Diener, E. (2001). Making the best of a bad situation: Satisfaction in the slums of Calcutta. *Social Indicators Research*, 55, 329–352.

Bjorklund, D. F. (1997). The role of immaturity in human development. *Psychological Bulletin*, 122, 153–169.

Blake, W. (1975/1790–1793). *The marriage of heaven and hell*. London: Oxford University Press.

Bloom, P. (2004). *Descartes' baby: How the science of child development explains what makes us human*. New York: Basic Books.

Blum, D. (2002). *Love at Goon Park*. Cambridge, MA: Perseus.

Boehm, C. (1999). *Hierarchy in the forest: The evolution of egalitarian behavior*. Cambridge, MA: Harvard University Press.

Boethius. (1962/c. 522 CE). *The consolation of philosophy*. (R. Green, Trans.). New York: Macmillan.

Bonanno, G. (2004). Loss, trauma, and human resilience: Have we underestimated the human capacity to thrive after extremely aversive events? *American Psychologist*, 59, 20–28.

Bouchard, T. J. (2004). Genetic influence on human psychological traits: A survey. *Current Directions in Psychological Science, 13,* 148–151.

Bowlby, J. (1969). *Attachment and loss.* Vol.1, *Attachment.* New York: Basic Books.

Boyer, P. (2001). *Religion explained: The evolutionary origins of religious thought.* New York: Basic Books.

Brickman, P., & Campbell, D. T. (1971). Hedonic relativism and planning the good society. In M. H. Apley (Ed.), *Adaptation-level theory: A symposium* (pp. 287–302). New York: Academic Press.

Brickman, P., Coates, D., & Janoff-Bulman, R. (1978). Lottery winners and accident victims: Is happiness relative? *Journal of Personality and Social Psychology, 36,* 917–927.

Brim, G. (1992). *Ambition.* New York: Basic Books.

Broderick, J. C. (Ed.). (1990). *Writings of Henry D. Thoreau: Journal, Volume 3: 1848–1851.* Princeton: Princeton University Press.

Bronte, C. (1973/1847). *Jane Eyre.* London: Oxford University Press.

Brown, D. E. (1991). *Human universals.* Philadelphia: Temple University Press.

Brown, R., & Gilman, A. (1960). The pronouns of power and solidarity. In T. A. Sebeok (Ed.), *Style in language* (pp. 253–276). Cambridge, MA: MIT Press.

Brown, S. L., Nesse, R. M., Vinokur, A. D., & Smith, D. M. (2003). Providing social support may be more beneficial than receiving it: Results from a prospective study of mortality. *Psychological Science, 14,* 320–327.

Buckley, T., & Gottlieb, A. (Eds.). (1988). *Blood magic: The anthropology of menstruation.* Berkeley: University of California Press.

Buchanan, D. C. (1965). *Japanese proverbs and sayings.* Norman, OK: University of Oklahoma Press.

Burns, D. D. (1999). *Feeling Good.* (2nd ed.). New York: Avon.

Burns, J. M., & Swerdlow, R. H. (2003). Right orbitofrontal tumor with pedophilia symptom and constructional apraxia sign. *Archives of Neurology, 60,* 437–440.

Bushman, B. J., & Baumeister, R. F. (1998). Threatened egotism, narcissism, self-esteem, and direct and displaced aggression: Does self-love or self-hate lead to violence? *Journal of Personality and Social Psychology, 75,* 219–229.

Buss, D. M. (2004). *Evolutionary psychology: The new science of the mind.* (2nd ed.). Boston: Allyn & Bacon.

Byrne, R., & Whiten, A. (Eds.). (1988). *Machiavellian intelligence.* Oxford, UK: Oxford University Press.

Byrom, T. (Ed. and Trans.). (1993). *Dhammapada: The sayings of the Buddha.* Boston: Shambhala.

Campbell, D. T. (1983). The two distinct routes beyond kin selection to ultrasociality: Implications for the humanities and social sciences. In D. Bridgeman (Ed.), *The nature of prococial development: Theories and strategies* (pp. 11–39). New York: Academic Press.

Carnegie, D. (1984/1944). *How to stop worrying and start living*. New York: Pocket Books.

Carstensen, L. L., Pasupathi, M., Mayr, U., & Nesselroade, J. R. (2000). Emotional experience in everyday life across the adult life span. *Journal of Personality and Social Psychology, 79*, 644–655.

Carter, C. (1998). Neuroendocrine perspectives on social attachment and love. *Psychoneuroendocrinology, 23*, 779–818.

Carver, C. S., & White, T. L. (1994). Behavioral inhibition, behavioral activation, and affective responses to impending reward and punishment: The BIS/BAS scales. *Journal of Personality and Social Psychology, 67*, 319–333.

Carver, C. S., Scheier, M. F., & Weintraub, J. K. (1989). Assessing coping strategies: A theoretically based approach. *Journal of Personality and Social Psychology, 56*, 267–283.

Cassidy, J. (1999). The nature of the child's ties. In J. Cassidy & P. R. Shaver (Eds.), *Handbook of attachment: Theory, research, and applications* (pp. 3–20). New York: Guilford.

Cather, W. (1987/1918). *My Antonia*; New York: Library of America.

Chan, W. T. (1963). *A source book in Chinese philosophy*. Princeton, NJ: Princeton University Press.

Chorpita, B. F., & Barlow, D. H. (1998). The development of anxiety: The role of control in the early environment. *Psychological Bulletin, 124*, 3–21.

Churchland, P. M. (1998). Toward a cognitive neuriobiology of the moral virtues. *Topoi, 17*, 83–96.

Cialdini, R. B. (2001). *Influence: Science and practice*. (4th ed.). Boston: Allyn and Bacon.

Cialdini, R. B., Vincent, J. E., Lewis, S. K., Catalan, J., Wheeler, D., & Darby, B. L. (1975). Reciprocal concessions procedure for inducing compliance: The door-in-the-face technique. *Journal of Personality and Social Psychology, 31*, 206–215.

Clark, A. (1999). *Being there: Putting brain, body, and world together again*. Cambridge, MA: MIT Press.

Cleckley, H. (1955). *The mask of sanity*. St. Louis, MO: Mosby.

Cohen, S., & Herbert, T. B. (1996). Health psychology: psychological factors and physical disease from the perspective of human psychoneuroimmunology. *Annual Reviews of Psychology, 47*, 113–142.

Conze, E. (Ed.). (1954). *Buddhist texts through the ages.* New York: Philosophical Library.

Conze, E. (Ed.). (1959). *Buddhist Scriptures.* London: Penguin.

Cooper, J. M. (Ed.). (1997). *Plato: Complete works.* Indianapolis, IN: Hackett.

Cosmides, L., & Tooby, J. (2004). Knowing thyself: The evolutionary psychology of moral reasoning and moral sentiments. *Business, Science, and Ethics,* 91–127.

Costa, P. T. J., & McCrae, R. R. (1989). Personality continuity and the changes of adult life. In M. Storandt & G. R. VandenBos (Eds.), *The adult years: Continuity and change* (pp. 45–77). Washington, DC: American Psychological Association.

Cross, P. (1977). Not can but will college teaching be improved. *New Directions for Higher Education,* 17, 1–15.

Cruikshank, B. (1999). *Will to empower: Democratic citizens and other subjects.* Ithaca: Cornell University Press.

Csikszentmihalyi, M. (1990). *Flow: The psychology of optimal experience.* New York: Harper & Row.

Csikszentmihalyi, M. (1997). *Finding flow.* New York: Basic Books.

Dalai Lama. (2001/1995). *The art of living: A guide to contentment, joy, and fulfillment.* (G. T. Jinpa, Trans.) London: Thorsons.

Damasio, A. (1994). *Descartes' error: Emotion, reason, and the human brain.* New York: Putnam.

Damasio, A. R., Tranel, D., & Damasio, H. (1990). Individuals with sociopathic behavior caused by frontal damage fail to respond autonomically to social stimuli. *Behavioral Brain Research,* 41, 81–94.

Damon, W. (1997). *The youth charter: How communities can work together to raise standards for all our children.* New York: Free Press.

Damon, W., Menon, J. & Bronk, K. (2003). The development of purpose during adolescence. *Applied Developmental Science* 7, 119–128.

Darwin, C. (1998/1871). *The descent of man and selection in relation to sex.* Amherst, NY: Prometheus.

Daston, L., and Park, C. (1998). *Wonders and the order of nature, 1150-1750.* New York: Zone.

Davidson, R. J. (1994). Asymmetric brain function, affective style, and psychopathology: The role of early experience and plasticity. *Development and Psychopathology,* 6, 741–758.

Davidson, R. J. (1998). Affective style and affective disorders: Perspectives from affective neuroscience. *Cognition and Emotion,* 12, 307–330.

Davidson, R. J., and Fox, N. A. (1989). Frontal brain asymmetry predicts infants' response to maternal separation. *Journal of Abnormal Psychology,* 98, 127–131.

Dawkins, R. (1976). *The selfish gene.* Oxford, UK: Oxford University Press.

DePaulo, B. M., & Morris, W. L. (2005). Singles in society and science. *Psychological Inquiry,* 16, 57–83.

DeRubeis, R. J., Hollon, S. D., Amsterdam, J. D., Shelton, R. C., Young, P. R., Salomon, R. M., et al. (2005). Cognitive therapy vs medications in the treatment of moderate to severe depression. *Archives of General Psychiatry,* 62, 409–416.

DeWolff, M., & van Ijzendoorn, M. (1997). Sensitivity and attachment: A meta-analysis on parental antecedents of infant attachment. *Child Development,* 68, 571–591.

Dharmakirti. (2002). *Mahayana tantra.* New Delhi, India: Penguin.

Diener, E., & Diener, C. (1996). Most people are happy. *Psychological Science,* 7, 181–185.

Diener, E., & Oishi, S. (2000). Money and happiness: Income and subjective well-being across nations. In E. Diener & E. M. Suh (Eds.), *Culture and subjective well-being* (pp. 185–218). Cambridge, MA: MIT Press.

Diener, E., & Suh, M. E. (1998). Subjective well-being and age: An international analysis. In K. Schaie & M. Lawton (Eds.), *Annual review of gerontology and geriatrics, Vol 17: Focus on emotion and adult development, Annual review of gerontology and geriatrics* (pp. 304–324). New York: Springer.

Diener, E., Suh, E. M., Lucas, R. E., & Smith, H. L. (1999). Subjective well-being: Three decades of progress. *Psychological Bulletin,* 125, 276–302.

Diener, E., Wolsic, B., & Fujita, F. (1995). Physical attractiveness and subjective well-being. *Journal of Personality and Social Psychology,* 69, 120–129.

Dijksterhuis, A., & van Knippenberg, A. (1998). The relation between perception and behavior, or how to win a game of Trivial Pursuit. *Journal of Personality and Social Psychology,* 74, 865–877.

Dobson, K. S. (1989). A meta-analysis of the efficacy of cognitive therapy for depression. *Journal of Consulting and Clinical Psychology,* 57, 414–419.

Doniger, W., & Smith, B. (Eds. & Trans.). (1991). *The laws of Manu.* London: Penguin.

Donne, J. (1975/1623). *Devotions upon emergent occasions: A critical edition with introduction and commentary.* Salzburg: University of Salzburg.

Donnellan, M. B., Trzesniewski, K. H., Robins, R. W., Moffitt, T. E., & Caspi, A. (2005). Low self-esteem is related to aggression, antisocial behavior, and delinquency. *Psychological Science, 16,* 328–335.

Dunbar, R. (1993). Coevolution of neocortical size, group size and language in humans. *Behavioral and Brain Sciences*, 16, 681–735.

Dunbar, R. (1996). *Grooming, gossip, and the evolution of language*. Cambridge, MA: Harvard University Press.

Dunning, D., Meyerowitz, J. A., & Holzberg, A. D. (2002). Ambiguity and self-evaluation: The role of idiosyncratic trait definitions in self-serving assessments of ability. In *Heuristics and biases: The psychology of intuitive judgment*. (pp. 324–333). Cambridge, UK: Cambridge University Press.

Durkheim, E. (1951/1897). *Suicide*. (J. A. Spalding & G. Simpson, Trans.) New York: Free Press.

Durkheim, E. (1965/1915). *The elementary forms of the religious life*. (J. W. Swain, Trans.) New York: Free Press.

Ekman, P., Sorensen, E., & Friesen, W. V. (1969). Pan-cultural elements in the facial displays of emotion. *Science*, 164, 86–88.

Elder, G. H., Jr. (1974). *Children of the great depression*. Chicago: University of Chicago Press.

Elder, G. H. (1998). The life course and human development. In R. M. Lerner (Ed.), *Handbook of child psychology*. Vol. 1, *Theoretical models of human development* (pp. 939–991). New York: Wiley.

Eliade, M. (1959/1957). *The sacred and the profane: The nature of religion*. (W. R. Task, Trans.). San Diego, CA: Harcourt Brace.

Emerson, R. W. (1960a/1838). The divinity school address. In S. Whicher (Ed.), *Selections from Ralph Waldo Emerson* (pp. 100–116). Boston: Houghton Mifflin.

Emerson, R. W. (1960b/1838). Nature. In S. Whicher (Ed.), *Selections from Ralph Waldo Emerson* (pp. 21–56). Boston: Houghton Mifflin.

Emmons, R. A. (1999). *The psychology of ultimate concerns: Motivation and spirituality in personality*. New York: Guilford.

Emmons, R. A. (2003). Personal goals, life meaning, and virtue: Wellsprings of a positive life. In C. L. M. Keyes & J. Haidt (Eds.), *Flourishing: Positive psychology and the life well-lived* (pp. 105–128). Washington DC: American Psychological Association.

Emmons, R. A., & McCullough, M. E. (2003). Counting blessings versus burdens: An experimental investigation of gratitude and subjective well-being in daily life. *Journal of Personality and Social Psychology*, 84, 377–389.

Epictetus (1983/1st–2nd cent. CE). *The manual*. (N. White, Trans.). Indianapolis, IN: Hackett.

Epicurus (1963/c. 290 BCE). *The philosophy of Epicurus*. (G. K. Strodach, Trans.). Chicago: Northwestern University Press.

Epley, N., & Caruso, E. M. (2004). Egocentric ethics. *Social Justice Research*, 17, 171–187.

Epley, N., & Dunning, D. (2000). Feeling "holier than thou": Are self-serving assessments produced by errors in self- or social prediction. *Journal of Personality and Social Psychology*, 79, 861–875.

Erikson, E. H. (1963/1950.) *Childhood and society.* (2nd ed.). New York: Norton.

Fadiman, J., & Frager, R. (Eds.). (1997). *Essential Sufism.* San Francisco: HarperSanFrancisco.

Fazio, R. H., Sanbonmatsu, D. M., Powell, M. C., & Kardes, F. R. (1986). On the automatic evaluation of attitudes. *Journal of Personality and Social Psychology*, 50, 229–238.

Feeney, J. A., & Noller, P. (1996). *Adult attachment.* Thousand Oaks, CA: Sage.

Feinberg, T. E. (2001). *Altered egos: How the brain creates the self.* New York: Oxford University Press.

Feingold, A. (1992). Good looking people are not what we think. *Psychological Bulletin*, 111, 304–341.

Feng, G. F., & English, J. (Eds.). (1972). *Tao Te Ching.* New York: Random House.

Fenton, T. (2005.) *Bad news: The decline of reporting, the business of news, and the danger to us all.* New York: Regan Books.

Fisher, H. (2004). *Why we love: The nature and chemistry of romantic love.* New York: Henry Holt.

Fitzgerald, J. M. (1988). Vivid memories and the reminiscence phenomenon: The role of a self-narrative. *Human Development*, 31, 261–273.

Flanagan, O. (1991.) *Varieties of moral personality: Ethics and psychological realism.* Cambridge, MA: Harvard University Press.

Fleeson, W., Malanos, A. B., & Achille, N. M. (2002). An intraindividual process approach to the relationship between extraversion and positive affect: Is acting extraverted as "good" as being extraverted? *Journal of Personality and Social Psychology*, 83, 1409–1422.

Frank, R. H. (1999). *Luxury fever: Why money fails to satisfy in an era of excess.* New York: Free Press.

Frank, R. H. (1988). *Passions within reason: The strategic role of the emotions.* New York: Norton.

Frankl, V. E. (1984). *Man's search for meaning.* New York: Pocket Books.

Franklin, B. (1962/c. 1791). *Autobiography of Benjamin Franklin.* New York: MacMillan.

Franklin, B. (1980/1733–1758). *Poor Richard's Almanack (selections).* Mount Vernon, NY: Peter Pauper Press.

Frederick, S., & Loewenstein, G. (1999). Hedonic adaptation. In D. Kahneman, E. Diener & N. Schwartz (Eds.), *Well-being: The foundations of hedonic psychology* (pp. 302–329). New York: Russell Sage.

Fredrickson, B. L. (1998). What good are positive emotions? *Review of General Psychology, 2,* 300–319.

Fredrickson, B. L. (2001). The role of positive emotions in positive psychology: The broaden-and-build theory of positive emotions. *American Psychologist, 56,* 218–226.

Freitas, A. L., and Higgins, E. T. (2002). Enjoying goal-directed action: The role of regulatory fit. *Psychological Science, 13,* 1–6.

Freud, S. (1976/1900). *The interpretation of dreams.* (J. Strachey, Trans.) New York: Norton.

Gallup, G. (1982). Self-awareness and the emergence of mind in primates. *American Journal of Primatology, 2,* 237–248.

Gardner, H., Csikszentmihalyi, M., & Damon, W. (2001). *Good work: When excellence and ethics meet.* New York: Basic Books.

Gazzaniga, M. S. (1985). *The social brain.* New York: Basic Books.

Gazzaniga, M. S., Bogen, J. E., & Sperry, R. W. (1962). Some functional effects of sectioning the cerebral commissures in man. *Proceedings of the National Academy of Sciences, USA, 48,* 1765–1769.

Geertz, C. (1973). Thick description: Toward an interpretive theory of culture. In C. Geertz (Ed.), *The interpretation of cultures.* New York: Basic Books.

Gershon, M. D. (1998). *The second brain.* New York: HarperCollins.

Gibbard, A. (1990). *Wise choices, apt feelings.* Cambridge, MA: Harvard University Press.

Gibran, K. (1977/1923). *The prophet.* New York: Alfred A. Knopf.

Gladwell, M. (2000). *The tipping point: How little things can make a big difference.* New York: Little Brown.

Gladwell, M. (2005). *Blink: The power of thinking without thinking.* New York: Little, Brown.

Glass, D. C., & Singer, J. E. (1972). *Urban stress; Experiments on noise and social stressors.* New York: Academic Press.

Glover, J. (2000). *Humanity: A moral history of the twentieth century.* New Haven, CT: Yale University Press.

Goldenberg, J. L., Pyszczynski, T., Greenberg, J., McCoy, S. K., & Solomon, S. (1999). Death, sex, love, and neuroticism: Why is sex such a problem? *Journal of Personality and Social Psychology, 77,* 1173–1187.

Goldenberg, J. L., Pyszczynski, T., Greenberg, J., Solomon, S., Kluck, B., & Cornwell, R. (2001). I am NOT an animal: Mortality salience, disgust, and the denial of human creatureliness. *Journal of Experimental Psychology: General*, 130, 427–435.

Gottman, J. (1994). *Why marriages succeed or fail*. New York: Simon & Schuster.

Graham, J., and Haidt, J. (manuscript in preparation). *The implicit and explicit moral values of liberals and conservatives*. University of Virginia, Dept. of Psychology.

Gray, J. A. (1994). Framework for a taxonomy of psychiatric disorder. In S. H. M. van Goozen & N. E. Van de Poll (Eds.), *Emotions: Essays on emotion theory* (pp. 29–59). Hillsdale, NJ: Lawrence Erlbaum.

Gray, J. G. (1970/1959). *The Warriors: Reflections of men in battle*. New York: Harper & Row.

Grob, C. S., & de Rios, M. D. (1994). Hallucinogens, managed states of consciousness, and adolescents: Cross-cultural perspectives. In P. K. Bock (Ed.), *Psychological Anthropology* (pp. 315–329). Westport, CT: Praeger.

Gross, J., & Haidt, J. (2005). The morality and politics of self-change. Unpublished manuscript, University of Virginia.

Guth, W., Schmittberger, R., and Schwarze, B. (1982). An experimental analysis of ultimatum bargaining. *Journal of Economic Behavior and Organization*, 3, 367–388.

Guyer, P. (Ed.). (1992). *The Cambridge companion to Kant*. Cambridge, UK: Cambridge University Press.

Haidt, J. (2001). The emotional dog and its rational tail: A social intuitionist approach to moral judgment. *Psychological Review, 108*, 814–834.

Haidt, J. (2003). Elevation and the positive psychology of morality. In C. L. M. Keyes and J. Haidt (Eds.), *Flourishing: Positive psychology and the life well-lived* (pp. 275–289). Washington, DC: American Psychological Association.

Haidt, J., Koller, S., and Dias, M. (1993). Affect, culture, and morality, or is it wrong to eat your dog? *Journal of Personality and Social Psychology, 65*, 613–628.

Haidt, J., Rozin, P., McCauley, C. R., & Imada, S. (1997). Body, psyche, and culture: The relationship between disgust and morality. *Psychology and Developing Societies, 9*, 107–131.

Haidt, J., & Rodin, J. (1999). Control and efficacy as interdisciplinary bridges. *Review of General Psychology, 3*, 317–337.

Haidt, J., & Hersh, M. A. (2001). Sexual morality: The cultures and reasons of liberals and conservatives. *Journal of Applied Social Psychology, 31*, 191–221.

Haidt, J., Rosenberg, E., & Hom, H. (2003). Differentiating diversities: Moral diversity is not like other kinds. *Journal of Applied Social Psychology, 33*, 1–36.

Haidt, J., & Joseph, C. (2004). Intuitive ethics: How innately prepared intuitions generate culturally variable virtues. *Daedalus* (Fall), 55–66.

Haidt, J., & Keltner, D. (2004). Appreciation of beauty and excellence. In C. Peterson and M. E. P. Seligman (Eds.), *Character strengths and virtues* (pp. 537–551). Washington, DC: American Psychological Association.

Haidt, J., & Bjorklund, F. (in press). Social intuitionists answer six questions about morality. In W. Sinnott-Armstrong (Ed.), *Moral psychology.* Vol. 2, *The cognitive science of morality.*

Hamer, D. H. (2004). *The God gene: How faith is hardwired into our genes.* New York: Doubleday.

Hamilton, W. D. (1964). The genetical evolution of social behavior, parts 1 and 2. *Journal of Theoretical Biology, 7,* 1–52.

Hansen, C. (1991). Classical Chinese Ethics. In P. Singer (Ed.), *A companion to ethics* (pp. 69–81). Oxford, UK: Basil Blackwell.

Haré, R. D. (1993). *Without conscience.* New York: Pocket Books.

Harker, L., & Keltner, D. (2001). Expressions of positive emotion in women's college yearbook pictures and their relationship to personality and life outcomes across adulthood. *Journal of Personality and Social Psychology, 80,* 112–124.

Harlow, H. F. (1971). *Learning to love.* San Francisco, CA: Albion.

Harlow, H. F., Harlow, M. K., & Meyer, D. R. (1950). Learning motivated by a manipulation drive. *Journal of Experimental Psychology, 40,* 228–234.

Harlow, H. F., & Zimmerman, R. (1959). Affectional responses in the infant monkey. *Science, 130,* 421–432.

Harris, J. R. (1995). Where is the child's environment? A group socialization theory of development. *Psychological Review, 102,* 458–489.

Hazan, C., & Shaver, P. (1987). Romantic love conceptualized as an attachment process. *Journal of Personality and Social Psychology, 52,* 511–524.

Hazan, C., & Zeifman, D. (1999). Pair bonds as attachments. In J. Cassidy & P. R. Shaver (Eds.), *Handbook of attachment: Theory, research, and applications* (pp. 336–354). New York: Guilford.

Heine, S. J., & Lehman, D. R. (1999). Culture, self-discrepancies, and self-satisfaction. *Personality and Social Psychology Bulletin, 25,* 915–925.

Helson, H. (1964). *Adaptation level theory: An experimental and systematic approach to behavior.* New York: Harper & Row.

Hick, J. (1967). The problem of evil. In P. Edwards (Ed.), *The Encyclopedia of Philosophy,* Vols. 3 & 4 (pp. 136–141). New York: Macmillan.

Hill, K., & Hurtado, A. M. (1996). *Ache life history.* New York: Aldine de Gruyter.

Hollon, S. D., & Beck, A. T. (1994). Cognitive and cognitive-behavioral therapies. In A. E. Bergin & S. L. Garfield (Eds.), *Handbook of psychotherapy and behavior change* (4th ed.). New York: Wiley.

Hollon, S. D., DeRubeis, R. J., Shelton, R. C., & Weiss, B. (2002). The emperor's new drugs: Effect size and moderation effects. *Prevention and Treatment,* 5, n.p.

Holmes, O. W., Jr. (1891). *Speeches.* Boston: Little, Brown.

Hom, H., & Haidt, J. (in preparation). The bonding and norming functions of gossip. Unpublished manuscript, University of Virginia.

Hoorens, V. (1993). Self-enhancement and superiority biases in social comparisons. In Vol. 4 of W. Strobe & M. Hewstone (Eds.), *European review of social psychology* (pp. 113–139). Chichester, UK: John Wiley.

Hume, D. (1969/1739). *A treatise of human nature.* London: Penguin.

Hunter, J. D. (2000). *The death of character: Moral education in an age without good and evil.* New York: Basic Books.

Irving, W. (1976). *George Washington: A biography.* Charles Neider (Ed.). Garden City, NY: Doubleday.

Isen, A. M., & Levin, P. F. (1972). Effect of feeling good on helping: Cookies and kindness. *Journal of Personality and Social Psychology,* 21, 384–388.

Isen, A. M., & Simmonds, S. (1978). The effect of feeling good on a helping task that is incompatible with good mood. *Social Psychology,* 41, 346–349.

Ito, T. A., & Cacioppo, J. T. (1999). The psychophysiology of utility appraisals. In D. Kahneman, E. Diener, and N. Schwarz (Eds.), *Well-being: The foundations of hedonic psychology* (pp. 470–488). New York: Russell Sage Foundation.

Iyengar, S. S., & Lepper, M. R. (2000). When choice is demotivating: Can one desire too much of a good thing? *Journal of Personality and Social Psychology,* 79, 995–1006.

James, W. (1950/1890). *The principles of psychology.* Vol. 2. New York: Dover.

James, W. (1961/1902). *The varieties of religious experience.* New York: Macmillan.

James, J. M., & Bolstein, R. (1992). Effect of monetary incentives and follow-up mailings on the response rate and response quality in mail surveys. *Public Opinion Quarterly,* 54, 442–453.

Jankowiak, W. R., & Fischer, E. F. (1992). A cross-cultural perspective on romantic love. *Ethnology,* 31, 149–155.

Jefferson, T. (1975/1771). Letter to Robert Skipwith. In M. D. Peterson (Ed.), *The portable Thomas Jefferson* (pp. 349–351). New York: Penguin.

Julien, R. M. (1998). *A primer of drug action.* (8th ed.). New York: W. H. Freeman.

Jung, C. G. (1963). *Memories, dreams, reflections.* New York: Pantheon.

Kagan, J. (1994). *Galen's prophecy: Temperament in human nature.* New York: Basic Books.

Kagan, J. (2003). Biology, context, and developmental inquiry. *Annual Review of Psychology,* 54, 1–23.

Kahneman, D., & Tversky, A. (1979). Prospect theory: An analysis of decisions under risk. *Econometrica, 47,* 263–291.

Kant, I. (1959/1785). *Foundation of the metaphysics of morals.* (L. W. Beck, Trans.) Indianapolis, IN: Bobbs-Merrill.

Kaplan, H. R. (1978). *Lottery winners: How they won and how winning changed their lives.* New York: Harper and Row.

Kass, L. R. (1994). *The hungry soul: Eating and the perfecting of our nature.* Chicago: University of Chicago.

Kasser, T. (2002). *The high price of materialism.* Cambridge, MA: MIT Press.

Kasser, T., & Ryan, R. M. (1996). Further examining the American dream: Differential correlates of intrinsic and extrinsic goals. *Personality and Social Psychology Bulletin, 22,* 280–287.

Keller, H. (1938). *Helen Keller's journal.* Garden City, NY: Doubleday.

Keltner, D., & Haidt, J. (2003). Approaching awe, a moral, spiritual, and aesthetic emotion. *Cognition and Emotion, 17,* 297–314.

Keyes, C. L. M., & Haidt, J. (Eds.). (2003). *Flourishing: Positive psychology and the life well lived.* Washington, DC: American Psychological Association.

King, L. A. (2001). The hard road to the good life: The happy, mature person. *Journal of Humanistic Psychology, 41,* 51–72.

Klemke, E. D. (Ed.). (2000). *The meaning of life.* (2nd ed.). New York: Oxford University Press.

Kohn, M. L., and Schooler, C. (1983). *Work and personality: An inquiry into the impact of social stratification.* Norwood, NJ: Ablex.

Kosfeld, M., Heinrichs, M., Zak, P. J., Fischbacher, U., & Fehr, E. (2005). Oxytocin increases trust in humans. *Nature, 435,* 673–676.

Koslowsky, M., & Kluger, A. N. (1995). *Commuting stress.* New York: Plenum.

Kramer, P. D. (1993). *Listening to Prozac.* New York: Viking.

Kuhn, D. (1991). *The skills of argument.* Cambridge, UK: Cambridge University Press.

Kunda, Z. (1990). The case for motivated reasoning. *Psychological Bulletin, 108,* 480–498.

Kunz, P. R., & Woolcott, M. (1976). Season's greetings: from my status to yours. *Social Science Research, 5,* 269–278.

LaBar, K. S., & LeDoux, J. E. (2003). Emotional learning circuits in animals and humans. In R. J. Davidson, K. R. Scherer & H. H. Goldsmith (Eds.), *Handbook of affective sciences* (pp. 52–65). Oxford, UK: Oxford University Press.

Laertius, D. (1925/3rd cent. CE). *Lives of eminent philosophers.* (R. D. Hicks, Trans.) London: Heinemann.

Lakin, J. L., and Chartrand, T. L. (2003). Using nonconscious behavioral mimicry to create affiliation and rapport. *Psychological Science*, 14, 334–339.

Lakoff, G., & Johnson, M. (1980). *Metaphors we live by.* Chicago: University of Chicago Press.

Lakoff, G., & Johnson, M. (1999). *Philosophy in the flesh.* New York: Basic Books.

Langer, E. J., & Rodin, J. (1976). The effects of choice and enhanced personal responsibility for the aged: A field experiment in an institutional setting. *Journal of Personality and Social Psychology*, 34, 191–198.

Lazarus, R. S., & Folkman, S. (1984). *Stress, appraisal, and coping.* New York: Springer.

Leakey, R. (1994). *The origin of humankind.* New York: Basic Books.

Leary, M. (2004). *The curse of the self: Self-awareness, egotism, and the quality of human life.* Oxford, UK: Oxford University Press.

Le Conte, J. (1892). *Evolution: Its nature, its evidences, and its relation to religious thought.* (2nd ed.). New York: D. Appleton.

LeDoux, J. (1996). *The Emotional Brain.* New York: Simon & Schuster.

Lerner, M. J., & Miller, D. T. (1978). Just world research and the attribution process: Looking back and ahead. *Psychological Bulletin*, 85, 1030–1051.

Leys, S. (Ed.). (1997). *The analects of Confucius.* New York: Norton.

Lichtheim, M. (1976). *Ancient egyptial literature: A book of readings.* Vol. 2, *The new kingdom.* Berkeley: University of California.

Lorenz, K. J. (1935). Der kumpan in der umwelt des vogels. *Journal für Ornithologie*, 83, 137–213.

Lucas, R. E. (2005). Happiness can change: A longitudinal study of adaptation to disability. Unpublished manuscript. Michigan State University.

Lucas, R. E., Clark, A. E., Georgellis, Y., & Diener, E. (2003). Reexamining adaptation and the set point model of happiness: Reactions to changes in marital status. *Journal of Personality and Social Psychology*, 84, 527–539.

Lucas, R. E., & Dyrenforth, P. S. (in press). Does the existence of social relationships matter for subjective well-being? In K. D. Vohs & E. J. Finkel (Eds.), *Intrapersonal processes and interpersonal relationships: Two halves, one self.* New York: Gulford.

Lucas, R. E., & Gohm, C. L. (2000). Age and sex differences in subjective well-being across cultures. In E. Diener & E. M. Suh (Eds.), *Culture and subjective well-being* (pp. 291–318). Cambridge, MA: MIT press.

Lucretius. (1977/c. 59 BCE). *The nature of things.* (F. O. Copley, Trans.) New York: Norton.

Lykken, D. T. (1999). *Happiness: What studies on twins show us about nature, nurture, and the happiness set-point.* New York: Golden Books.

Lykken, D. T., McGue, M., Tellegen, A., & Bouchard, T. J. (1992). Emergenesis: Genetic traits that may not run in families. *American Psychologist, 47*, 1565–1577.

Lykken, D. T., & Tellegen, A. (1996). Happiness is a stochastic phenomenon. *Psychological Science, 7*, 186–189.

Lynn, M., & McCall, M. (1998). Beyond gratitude and gratuity. Unpublished manuscript, Cornell University, School of Hotel Administration, Ithaca, NY.

Lyte, M., Varcoe, J. J., & Bailey, M. T. (1998). Anxiogenic effect of subclinical bacterial infection in mice in the absence of overt immune activation. *Physiology and behavior, 65*, 63–68.

Lyubomirsky, S., King, L., & Diener, E. (in press). The benefits of frequent positive affect: Does happiness lead to success? *Psychological Bulletin*.

Lyubomirsky, S., Sheldon, K. M., & Schkade, D. (in press). Pursuing happiness: The architecture of sustainable change. *Review of General Psychology*.

Machiavelli, N. (1940/c. 1517). *The prince and the discourses*. (L. Ricci & C. E. Detmold, Trans.). New York: Modern Library.

MacIntyre, A. (1981). *After virtue*. Notre Dame, IN: University of Notre Dame Press.

Marcus, G. (2004). *The birth of the mind*. New York: Basic Books.

Margolis, H. (1987). *Patterns, thinking, and cognition*. Chicago: University of Chicago Press.

Markus, H. R., & Kitayama, S. (1991). Culture and the self: Implications for cognition, emotion, and motivation. *Psychological Review, 98*, 224–253.

Marx, K. (1977/1867). *Capital: A critique of political economy*. New York: Vintage.

Mascaro, J. (Ed. and Trans.). (1965). *The Upanishads*. London: Penguin.

Mascaro, J. (Ed. and Trans.). (1973). *The Dhammapada*. Harmondsworth, UK: Penguin.

Maslow, A. H. (1964). *Religions, values, and peak-experiences*. Columbus, OH: Ohio State University Press.

Mastekaasa, A. (1994). Marital status, distress, and well-being: An international comparison. *Journal of Comparative Family Studies, 25*, 183–205.

McAdams, D. P. (1994). Can personality change? Levels of stability and growth in personality across the life span. In T. F. Heatherton & J. L. Weinberger (Eds.), *Can personality change?* (pp. 299–313). Washington, DC: American Psychological Association.

McAdams, D. P. (2001). The psychology of life stories. *Review of General Psychology, 5*, 100–122.

McCraty, R., and Childre, D. (2004). The grateful heart: The psychophysiology of appreciation. In R. A. Emmons and M. E. McCullough (Eds.), *The psychology of gratitude* (pp. 230–255). New York: Oxford.

McCullough, M. E., Hoyt, W. T., Larson, D. B., Koenig, H. G., & Thoresen, C. (2000). Religious involvement and mortality: A meta-analytic review. *Health Psychology,* 1, 211–222.

McNeill, W. H. (1995). *Keeping together in time: Dance and drill in human history.* Cambridge, MA.: Harvard University Press.

Meichenbaum, D. (1985). *Stress innoculation training.* New York: Pergamon.

Metcalfe, J., & Mischel, W. (1999). A hot/cool-system analysis of delay of gratification: Dynamics of willpower. *Psychological Review,* 106, 3–19.

Miller, N. E. (1944). Experimental studies of conflict. In J. M. Hunt (Ed.), *Personality and the behavior disorders.* New York: Ronald Press.

Miller, W. I. (1997). *The anatomy of disgust.* Cambridge, MA: Harvard University Press.

Miller, W. R., & C'de Baca, J. (2001). *Quantum Change.* New York: Guilford.

Mithen, S. (2000). Mind, brain and material culture: An archaeological perspective. In P. Carruthers and A. Chamberlain (Eds.), *Evolution and the human mind* (pp. 207–217), Cambridge: Cambridge University Press.

Montaigne, M. (1991/1588). *The complete essays.* (M. A. Screech, Ed. & Trans.). London: Penguin.

Moss, C. (1998). *Elephant Memories: Thirteen years in the life of an elephant family.* New York: William Morrow.

Mroczek, D. K., & Spiro, A. (2005). Change in life satisfaction during adulthood: Findings from the veterans affairs normative aging study. *Journal of Personality and Social Psychology,* 88, 189–202.

Myers, D. G. (2000). The funds, friends, and faith of happy people. *American Psychologist,* 55, 56–67.

Nakamura, J., and Csikszentmihalyi, M. (2003). The construction of meaning through vital engagement. In C. L. M. Keyes and J. Haidt (Eds.), *Flourishing: Positive psychology and the life well-lived* (pp. 83–104). Washington, DC: American Psychological Association.

Nestler, E. J., Hyman, S. E., & Malenka, R. C. (2001). *Molecular neuropharmacology: A foundation for clinical neuroscience.* New York: McGraw-Hill.

Newberg, A., D'Aquili, E., & Rause, V. (2001). *Why God won't go away: Brain science and the biology of belief.* New York: Ballantine.

Nietzsche, F. (1997/1889). *Twilight of the idols.* (R. Polt, Trans.) Indianapolis, IN: Hackett.

Nolen-Hoeksema, S., & Davis, C. G. (2002). Positive responses to loss. In C. R. Snyder & S. J. Lopez (Eds.), *Handbook of positive psychology* (pp. 598–607). New York: Oxford.

Nosek, B. A., Banaji, M. R., & Greenwald, A. G. (2002). Harvesting intergroup implicit attitudes and beliefs from a demonstration web site. *Group Dynamics*, 6, 101–115.

Nosek, B. A., Greenwald, A. G., & Banaji, M. R. (in press). The Implicit Association Test at age 7: A methodological and conceptual review. In J. A. Bargh (Ed.), *Automatic processes in social thinking and behavior*. Philadelphia: Psychology Press.

Nussbaum, M. C. (2001). *Upheavals of thought*. Cambridge, UK: Cambridge University Press.

O'Connor, E. (Ed. & Trans.). (1993). *The essential Epicurus*. Amherst, NY: Prometheus Books.

Obeyesekere, G. (1985). Depression, Buddhism, and work of culture in Sri Lanka. In A. Klineman & B. Good (Eds.), *Culture and depression* (pp. 134–152). Berkeley: University of California Press.

Olds, J., & Milner, P. (1954). Positive reinforcement produced by electrical stimulation of septal areas and other regions of rat brains. *Journal of Comparative and Physiological Psychology*, 47, 419–427.

Ovid (2004/c. 10 CE). *Metamorphoses*. (D. Raeburn, Trans.). London: Penguin.

Pachocinski, R. (1996). *Proverbs of Africa: Human nature in the Nigerian oral tradition*. St. Paul, MN: Professors World Peace Academy.

Pahnke, W. N. (1966). Drugs and mysticism. *International Journal of Parapsychology*, 8, 295–313.

Panthanathan, K., and Boyd, R. (2004). Indirect reciprocity can stabilize cooperation without the second-order free rider problem. *Nature*, 432, 499–502.

Park, C. L., Cohen, L., & Murch, R. (1996). Assessment and prediction of stress-related growth. *Journal of Personality*, 64, 71–105.

Pelham, B. W., Mirenberg, M. C., & Jones, J. K. (2002). Why Susie sells seashells by the seashore: Implicit egotism and major life decisions. *Journal of Personality and Social Psychology*, 82, 469–487.

Pennebaker, J. (1997). *Opening up: The healing power of expressing emotions* (Rev. ed.). New York: Guilford.

Perkins, D. N., Farady, M., & Bushey, B. (1991). Everyday reasoning and the roots of intelligence. In J. F. Voss, D. N. Perkins & J. W. Segal (Eds.), *Informal reasoning and education* (pp. 83–105). Hillsdale, NJ: Erlbaum.

Peterson, C., & Seligman, M. E. P. (2004). *Character strengths and virtues: A handbook and classification.* Washington, DC: American Psychological Association and Oxford University Press.

Piaget, J. (1965/1932). *The moral judgment of the child.* (M. Gabain, Trans.) New York: Free Press.

Piliavin, J. A. (2003). Doing well by doing good: Benefits for the benefactor. In C. L. M. Keyes and J. Haidt (Eds.), *Flourishing: Positive psychology and the life well-lived* (pp. 227–247). Washington, DC: American Psychological Association.

Pincoffs, E. L. (1986). *Quandaries and virtues: Against reductivism in ethics.* Lawrence, KS: University of Kansas.

Pinker, S. (1997). *How the mind works.* New York: Norton.

Pinker, S. (2002). *The blank slate: The modern denial of human nature.* New York: Viking.

Plomin, R., & Daniels, D. (1987). Why are children in the same family so different from one another? *Behavioral and Brain Sciences,* 10, 1–60.

Pronin, E., Lin, D. Y., & Ross, L. (2002). The bias blind spot: Perceptions of bias in self versus others. *Personality and Social Psychology Bulletin,* 28, 369–381.

Proust, M. (1992a/1922). *In search of lost time.* Vol. 2, *Within a budding grove.* (C. K. S. Moncreiff & T. Kilmartin, Trans.) London: Chatto and Windus.

Proust, M. (1992b/1922). *In search of lost time.* Vol. 5, *The captive and the fugitive.* (C. K. S. Moncreiff & T. Kilmartin, Trans.) London: Chatto and Windus.

Putnam, R. D. (2000). *Bowling alone: The collapse and revival of American community.* New York: Simon & Schuster.

Pyszcsynski, T., Greenberg, J., & Solomon, S. (1997). Why do we want what we want? A terror management perspective on the roots of human social motivation. *Psychological Inquiry,* 8, 1–20.

Pyszczynski, T., & Greenberg, J. (1987). Toward an integration of cognitive and motivational perspectives on social inference: A biased hypothesis-testing model. *Advances in Experimental Social Psychology,* 20, 297–340.

Reis, H. T., & Gable, S. L. (2003). Toward a positive psychology of relationships. In C. L. M. Keyes & J. Haidt (Eds.), *Flourishing: Positive psychology and the life well-lived* (pp. 129–159). Washington, DC: American Psychological Association.

Richerson, P. J., & Boyd, R. (1998). The evolution of human ultra-sociality. In I. Eibl-Eibesfeldt & F. K. Salter (Eds.), *Indoctrinability, ideology, and warfare: Evolutionary perspectives* (pp. 71–95). New York: Berghahn.

Richerson, P. J., & Boyd, R. (2005). *Not by genes alone: How culture transformed human evolution.* Chicago: University of Chicago Press.

Ridley, M. (1996). *The origins of virtue.* Harmondsworth, UK: Penguin.

Riis, J., Loewenstein, G., Baron, J., Jepson, C., Fagerlin, A., & Ubel, P. A. (2005). Ignorance of hedonic adaptation to hemodialysis: A study using ecological momentary assessment. *Journal of Experimental Psychology: General,* 134, 3–9.

Rind, B., Tromovitch, P., & Bauserman, R. (1998). A meta-analytic examination of assumed properties of child sexual abuse using college samples. *Psychological Bulletin,* 124, 22–53.

Rodin, J., & Langer, E. (1977). Long-term effects of a control-relevant intervention with the institutionalized aged. *Journal of Personality and Social Psychology,* 35, 897–902.

Rolls, E. T. (1999). *The brain and emotion.* Oxford, UK: Oxford University Press.

Ross, M., & Sicoly, F. (1979). Egocentric biases in availability and attribution. *Journal of Personality and Social Psychology,* 37, 322–336.

Rozin, P., & Fallon, A. (1987). A perspective on disgust. *Psychological Review,* 94, 23–41.

Rozin, P., Haidt, J., McCauley, C., & Imada, S. (1997). Disgust: Preadaptation and the evolution of a food-based emotion. In H. MacBeth (Ed.), *Food preferences and taste* (pp. 65–82). Providence, RI: Berghahn.

Rozin, P., Haidt, J., & McCauley, C. (2000). Disgust. In M. Lewis and J. M. Haviland-Jones (Eds.), *Handbook of emotions* (pp. 637–653). New York: Guilford Press.

Rozin, P., and Royzman, E. B. (2001). Negativity bias, negativity dominance, and contagion. *Personality and Social Psychology Review,* 5, 296–320.

Russell, J. B. (1988). *The prince of darkness: Radical evil and the power of good in history.* Ithaca, NY: Cornell University Press.

Ryan, R. M., and Deci, E. L. (2000). Self-determination theory and the facilitation of intrinsic motivation, social development, and well-being. *American Psychologist,* 55, 68–78.

Ryff, C. D., & Singer, B. (2003). Flourishing under fire: Resilience as a prototype of challenged thriving. In C. L. M. Keyes & J. Haidt (Eds.), *Flourishing: Positive psychology and the life well-lived* (pp. 15–36). Washington, DC: American Psychological Association.

Sabini, J., & Silver, M. (1982). *Moralities of everyday life.* Oxford, UK: Oxford University Press.

Salovey, P., & Mayer, J. D. (1990). Emotional intelligence. *Imagination, Cognition, and personality,* 9, 185–211.

Sampson, R. J. (1993). Family management and child development: Insights from social disorganization theory. Vol. 6 of J. McCord (Ed.), *Advances in criminological theory* (pp. 63–93). New Brunswick, NJ: Transaction Press.

Sanfey, A. G., Rilling, J. K., Aronson, J. A., Nystrom, L. E., & Cohen, J. D. (2003). The neural basis of economic decision-making in the ultimatum game. *Science, 300,* 1755–1758.

Sartre, J. P. (1989/1944). *No exit and three other plays.* (S. Gilbert, Trans.). New York: Vintage International.

Schatzberg, A. F., Cole, J. O., & DeBattista, C. (2003). *Manual of Clinical Psychopharmacology,* (4th Ed.). Washington, DC: American Psychiatric Publishing.

Schkade, D. A., & Kahneman, D. (1998). Does living in California make people happy? A focusing illusion in judgments of life satisfaction. *Psychological Science, 9,* 340–346.

Schulz, R., & Decker, S. (1985). Long-term adjustment to physical disability: The role of social support, perceived control, and self-blame. *Journal of Personality and Social Psychology, 48,* 1162–1172.

Schwartz, B. (2004). *The paradox of choice.* New York: HarperCollins.

Schwartz, B., Ward, A., Monterosso, J., Lyubomirsky, S., White, K., & Lehman, D. R. (2002). Maximizing versus satisficing: Happiness is a matter of choice. *Journal of Personality and Social Psychology, 83,* 1178–1197.

Seligman, M. E. P. (1995). The effectiveness of psychotherapy: The Consumer Reports study. *American Psychologist, 50,* 965–974.

Seligman, M. E. P. (2002). *Authentic happiness.* New York: Free Press.

Seneca, L. A. (1917–1925/c. 50 CE). *Moral epistles.* Vol. 1, The Loeb Classical Library. Cambridge, MA: Harvard University Press.

Shapiro, S., Schwartz, G. E. R., & Santerre, C. (2002). Meditation and positive psychology. In C. R. Snyder & S. J. Lopez (Eds.), *Handbook of positive psychology* (pp. 632–645). New York: Oxford University Press.

Sheldon, K. M. (2004). *Optimal human being: An integrated multi-level perspective.* Mahwah, NJ: Lawrence Erlbaum.

Sheldon, K. M., & Kasser, T. (1995). Coherence and congruence: Two aspects of personality integration. *Journal of Personality and Social Psychology, 68,* 531–543.

Shoda, Y., Mischel, W., & Peake, P. K. (1990). Predicting adolescent cognitive and self-regulatory competencies from preschool delay of gratification: Identifying diagnostic conditions. *Developmental Psychology, 26,* 978–986.

Shore, B. (1996). *Culture in mind: Cognition, culture, and the problem of meaning.* New York: Oxford University Press.

Shulgin, A. (1991). *PIHKAL: A chemical love story.* Berkeley: Transform Press.

Shweder, R. A., Much, N. C., Mahapatra, M., & Park, L. (1997). The "big three" of morality (autonomy, community, and divinity), and the "big three" explanations of suffering. In A. Brandt & P. Rozin (Eds.), *Morality and Health* (pp. 119–169). New York: Routledge.

Singer, P. (1979). *Practical ethics.* Cambridge, UK: Cambridge University Press.

Skitka, L. J. (2002). Do the means always justify the ends, or do the ends sometimes justify the means? A value protection model of justice reasoning. *Personality and Social Psychology Bulletin, 28,* 588–597.

Smith, A. (1976/1759). *The theory of moral sentiments.* Oxford, UK: Oxford University Press.

Smith, N. M., Floyd, M. R., Scogin, F., & Jamison, C. S. (1997). Three year followup of bibliotherapy for depression. *Journal of Consulting and Clinical Psychology, 65,* 324–327.

Solnick, S. J., & Memenway, D. (1998). Is more always better? A survey on positional concerns. *Journal of Economic Behavior and Organization, 37,* 373–383.

Solomon, R. C. (1999). *The joy of philosophy: Thinking thin versus the passionate life.* New York: Oxford University Press.

Srivastava, S., John, O. P., Gosling, S. D., & Potter, J. (2003). Development of personality in early and middle addulthood: Set like plaster or persistent change? *Journal of Personality and Social Psychology, 84,* 1041–1053.

Stall, S. (1904/1897). *What a young man ought to know.* London: Vir Publishing.

Steele, J. D. (1867). *Fourteen weeks in chemistry.* New York: A. S. Barnes.

Sternberg, R. J. (1986). A triangular theory of love. *Psychological Review, 93,* 119–135.

Sternberg, R. J. (1998). A balance theory of wisdom. *Review of General Psychology, 2,* 347–365.

Tajfel, H. (1982). Social psychology of intergroup relations. *Annual Review of Psychology, 33,* 1–39.

Tamir, M., Robinson, M. D., & Clore, G. L. (2002). The epistemic benefits of trait-consistent mood states: An analysis of extraversion and mood. *Journal of Personality and Social Psychology, 83*(3), 663–677.

Tavris, C. (1982). *Anger: The misunderstood emotion.* New York: Simon & Schuster.

Taylor, C. (1989). *Sources of the self: The making of the modern identity.* Cambridge, MA: Harvard University Press.

Taylor, S. E. (2003). *Health psychology.* Boston: McGraw-Hill.

Taylor, S. E., Klein, L. C., Lewis, B. P., Gruenewald, T. L., Gurung, R. A., & Updegraff, J. A. (2000). Biobehavioral responses to stress in females: Tend-and-befriend, not fight-or-flight. *Psychological Review, 107,* 411–429.

Taylor, S. E., Lerner, J. S., Sherman, D. K., Sage, R. M., & McDowell, N. K. (2003). Portrait of the self-enhancer: Well adjusted and well liked or maladjusted and friendless. *Journal of Personality and Social Psychology*, 84, 165–176.

Tedeschi, R. G., Park, C. L., & Calhoun, L. G. (1998). Posttraumatic growth: Conceptual issues. In R. G. Tedeschi, C. L. Park & L. G. Calhoun (Eds.), *Posttraumatic growth: Positive changes in the aftermath of crisis* (pp. 1–22). Mahwah, NJ: Lawrence Erlbaum.

Templeton, J. M. (1997). *Worldwide laws of life: 200 eternal spiritual principles*. Philadelphia: Templeton Foundation Press.

Tennen, H., & Affleck, G. (1998). Personality and transformation in the face of adversity. In R. G. Tedeschi, C. L. Park & L. G. Calhoun (Eds.), *Posttraumatic growth: Positive changes in the aftermath of crisis* (pp. 65–98). Mahwah, NJ: Lawrence Erlbaum.

Thoits, P. A., & Hewitt, L. N. (2001). Volunteer work and well-being. *Journal of Health and Social Behavior*, 42, 115–131.

Thrash, T. M., and Elliot, A. J. (2004). Inspiration: Core characteristics, component processes, antecedents, and function. *Journal of Personality and Social Psychology*, 87, 957.

Thomas, K. (1983). *Man and the Natural World*. New York: Pantheon.

Tooby, J., & Cosmides, L. (1996). Friendship and the banker's paradox: Other pathways to the evolution of adaptations for altruism. *Proceedings of the British Academy*, 88, 119–143.

Trevathan, W. (1987). *Human birth*. New York: Aldine de Gruyter.

Trivers, R. L. (1971). The evolution of reciprocal altruism. *Quarterly Review of Biology*, 46, 35–57.

Troyat, H. (1967). *Tolstoy*. (N. Amphoux, Trans). New York: Doubleday.

Trut, L. N. (1999). Early canid domestication: The farm fox experiment. *American Scientist*, 87, 160–169.

Turkheimer, E. (2000). Three laws of behavior genetics and what they mean. *Current Directions in Psychological Science*, 9, 160–164.

Updegraff, J. A., & Taylor, S. E. (2000). From vulnerability to growth: Positive and negative effects of stressful life events. In J. Harvey & E. Miller (Eds.), *Loss and trauma: General and close relationship perspectives* (pp. 3–28). Philadelphia: Brunner-Routledge.

Uvnas-Moberg, K. (1998). Oxytocin may mediate the benefits of positive social interaction and emotions. *Psychoneuroimmunology*, 23, 819–835.

van Baaren, R. B., Holland, R. W., Steenaert, B., & van Knippenberg, A. (2003). Mimicry for money: Behavioral consequences of imitation. *Journal of Experimental Social Psychology*, 39, 393–398.

van Baaren, R. B., Holland, R. W., Kawakami, K., & van Knippenberg, A. (2004). Mimicry and Prosocial Behavior. *Psychological Science,* 15, 71–74.

Van Boven, L., & Gilovich, T. (2003). To do or to have? That is the question. *Journal of Personality and Social Psychology* 85, 1193–1202.

van IJzendoorn, M. H., Moran, G., Belsky, J., Pederson, D., Bakermans-Kranenburg, M. J., & Kneppers, K. (2000). The similarity of siblings' attachments to their mother. *Child Development,* 71, 1086–1098.

Vormbrock, J. K. (1993). Attachment theory as applied to war-time and job-related marital separation. *Psychological Bulletin,* 114, 122–144.

Waite, L. J., & Gallagher, M. (2000). *The case for marriage: Why married people are happier, healthier, and better off financially.* New York: Doubleday.

Warren, R. (2002). *The purpose driven life: What on earth am I here for?* Grand Rapids, MI: Zondervan.

Wasson, R. G. (1986). *Persephone's quest: Entheogens and the origins of religion.* New Haven, CT: Yale University Press.

Watson, J. B. (1928). *Psychological care of infant and child.* New York: W. W. Norton.

Wegner, D. (1994). Ironic processes of mental control. *Psychological Review,* 101, 34–52.

Weinfield, N. S., Sroufe, L. A., Egeland, B., & Carlson, E. A. (1999). The nature of individual differences in infant-caregiver attachment. In J. Cassidy & P. R. Shaver (Eds.), *Handbook of attachment: Theory, research, and applications* (pp. 68–88). New York: Guilford.

Wesley, J. (1986/1786). *Works of John Wesley.* A. Outler (Ed.). Nashville, TN: Abingdon Press.

White, R. B. (1959). Motivation reconsidered: The concept of competence. *Psychological Review* 66, 297–333.

Whybrow, P. C. (2005). *American mania: When more is not enough.* New York: Norton.

Wilkinson, G. S. (1984). Reciprocal food sharing in the vampire bat. *Nature,* 308, 181–184.

Williams, G. C. (1966). *Adaptation and natural selection: A critique of some current evolutionary thought.* Princeton: Princeton University Press.

Wilson, D. S. (2002). *Darwin's cathedral: Evolution, religion, and the nature of society.* Chicago: University of Chicago Press.

Wilson, E. O. (1990). *Success and dominance in ecosystems: The case of the social insects.* Oldendorf, Germany: Ecology Institute.

Wilson, T. D., & Gilbert, D. T. (2003). Affective forecasting. In Vol. 35 of M. P. Zanna (Ed.), *Advances in experimental psychology* (pp. 345–411). San Diego, CA: Academic.

Wilson, T. D., & Gilbert, D. T. (2005). Making sense: A model of affective adaptation. Unpublished manuscript.

Wright, R. (1994). *The moral animal.* New York: Pantheon.

Wrzesniewski, A., McCauley, C. R., Rozin, P., & Schwartz, B. (1997). Jobs, careers, and callings: People's relations to their work. *Journal of Research in Personality,* 31, 21–33.

Wrzesniewski, A., Rozin, P., & Bennett, G. (2003). Working, playing, and eating: Making the most of most moments. In C. L. M. Keyes & J. Haidt (Eds.), *Flourishing: Positive psychology and the life well-lived* (pp. 185–204). Washington, DC: American Psychological Association.

Zaehner, R. C. (Ed. and Trans.). (1969). *The Bhagavad-Gita.* Oxford: Clarendon.

# Index

Jonathan Haidt is the Thomas Cooley Professor of Ethical Leadership at New York University's Stern School of Business. His research examines the emotional basis of morality and the ways that morality varies across cultures, including the cultures of liberals and conservatives. He is the co-editor of *Flourishing: Positive Psychology and the Life Well-Lived*.

For further information on topics discussed in this book, visit www.happinesshypothesis.com.